NATIONAL ACADEMIE

Confucius Institutes at U.S. Institutions of Higher Education

Waiver Criteria for the Department of Defense

Philip J. Hanlon, Jayathi Y. Murthy, and Sarah M. Rovito, *Editors*

U.S. Science and Innovation Policy

Policy and Global Affairs

Consensus Study Report

NATIONAL ACADEMIES PRESS 500 Fifth Street, NW Washington, DC 20001

This activity was supported by a contract between the National Academy of Sciences and the U.S. Department of Defense. Any opinions, findings, conclusions, or recommendations expressed in this publication do not necessarily reflect the views of any organization or agency that provided support for the project.

International Standard Book Number-13: 978-0-309-69486-5
International Standard Book Number-10: 0-309-69486-8
Digital Object Identifier: https://doi.org/10.17226/26747

This publication is available from the National Academies Press, 500 Fifth Street, NW, Keck 360, Washington, DC 20001; (800) 624-6242 or (202) 334-3313; http://www.nap.edu.

Copyright 2023 by the National Academy of Sciences. National Academies of Sciences, Engineering, and Medicine and National Academies Press and the graphical logos for each are all trademarks of the National Academy of Sciences. All rights reserved.

Printed in the United States of America.

Suggested citation: National Academies of Sciences, Engineering, and Medicine. 2023. *Confucius Institutes at U.S. Institutions of Higher Education: Waiver Criteria for the Department of Defense*. Washington, DC: The National Academies Press. https://doi.org/10.17226/26747.

The **National Academy of Sciences** was established in 1863 by an Act of Congress, signed by President Lincoln, as a private, nongovernmental institution to advise the nation on issues related to science and technology. Members are elected by their peers for outstanding contributions to research. Dr. Marcia McNutt is president.

The **National Academy of Engineering** was established in 1964 under the charter of the National Academy of Sciences to bring the practices of engineering to advising the nation. Members are elected by their peers for extraordinary contributions to engineering. Dr. John L. Anderson is president.

The **National Academy of Medicine** (formerly the Institute of Medicine) was established in 1970 under the charter of the National Academy of Sciences to advise the nation on medical and health issues. Members are elected by their peers for distinguished contributions to medicine and health. Dr. Victor J. Dzau is president.

The three Academies work together as the **National Academies of Sciences, Engineering, and Medicine** to provide independent, objective analysis and advice to the nation and conduct other activities to solve complex problems and inform public policy decisions. The National Academies also encourage education and research, recognize outstanding contributions to knowledge, and increase public understanding in matters of science, engineering, and medicine.

Learn more about the National Academies of Sciences, Engineering, and Medicine at **www.nationalacademies.org**.

Consensus Study Reports published by the National Academies of Sciences, Engineering, and Medicine document the evidence-based consensus on the study's statement of task by an authoring committee of experts. Reports typically include findings, conclusions, and recommendations based on information gathered by the committee and the committee's deliberations. Each report has been subjected to a rigorous and independent peer-review process and it represents the position of the National Academies on the statement of task.

Proceedings published by the National Academies of Sciences, Engineering, and Medicine chronicle the presentations and discussions at a workshop, symposium, or other event convened by the National Academies. The statements and opinions contained in proceedings are those of the participants and are not endorsed by other participants, the planning committee, or the National Academies.

Rapid Expert Consultations published by the National Academies of Sciences, Engineering, and Medicine are authored by subject-matter experts on narrowly focused topics that can be supported by a body of evidence. The discussions contained in rapid expert consultations are considered those of the authors and do not contain policy recommendations. Rapid expert consultations are reviewed by the institution before release.

For information about other products and activities of the National Academies, please visit www.nationalacademies.org/about/whatwedo.

COMMITTEE ON CONFUCIUS INSTITUTES AT U.S. INSTITUTIONS OF HIGHER EDUCATION

PHILIP J. HANLON (*Chair*), President and Professor of Mathematics, Dartmouth College
JAYATHI Y. MURTHY (*Vice Chair*) [NAE],* President and Professor of Mechanical, Industrial and Manufacturing Engineering, Oregon State University
HANNAH L. BUXBAUM, Vice President for International Affairs and Professor of Law and John E. Schiller Chair, Indiana University
CLAUDE R. CANIZARES [NAS], Bruno Rossi Professor of Physics, Massachusetts Institute of Technology
ROBERT L. DALY, Director, Kissinger Institute on China and the United States, Woodrow Wilson International Center for Scholars
PETER K. DORHOUT, Vice President for Research and Professor of Chemistry, Iowa State University
MELISSA L. FLAGG, Founder, Flagg Consulting LLC
MARY GALLAGHER, Lowenstein Professor of Democracy, Democratization, and Human Rights, University of Michigan
JENNY J. LEE, Professor in the Department of Educational Policy Studies and Practice, University of Arizona
IVETT A. LEYVA, College of Engineering Excellence Professor and Department Head of Aerospace Engineering, Texas A&M University
ELIZABETH D. PELOSO, Associate Vice President and Associate Vice Provost of Research Services, University of Pennsylvania
JEFFREY M. RIEDINGER, Vice Provost of Global Affairs and Professor of Law, University of Washington
C. REYNOLD VERRET, President and Professor of Biochemistry, Xavier University of Louisiana

Study Staff

SARAH M. ROVITO, Study Director and Senior Program Officer, U.S. Science and Innovation Policy
TOM WANG, Policy Theme Lead and Senior Board Director, U.S. Science and Innovation Policy
FRAZIER F. BENYA, Senior Program Officer, U.S. Science and Innovation Policy
ANITA EISENSTADT, Program Officer, U.S. Science and Innovation Policy (through May 2022)

---

* Designates membership in the National Academy of Sciences (NAS), National Academy of Engineering (NAE), or National Academy of Medicine (NAM).

JOHN VERAS, Research Associate, U.S. Science and Innovation Policy
LESLEY SNYDER, Senior Program Assistant, U.S. Science and Innovation Policy (April to August 2022)
CLARA HARVEY-SAVAGE, Senior Finance Business Partner

Consultants

JOE ALPER, Consulting Writer
KORANTEMA KALEEM, Researcher, American Institutes for Research
JASMINE HOWARD, Qualitative Research Associate, American Institutes for Research
KELLIE MACDONALD, Research Associate, American Institutes for Research

Acknowledgment of Reviewers

This Consensus Study Report was reviewed in draft form by individuals chosen for their diverse perspectives and technical expertise. This independent review provides candid and critical comments that will assist the National Academies of Sciences, Engineering, and Medicine in making each published report as sound as possible and to ensure that it meets the institutional standards for quality, objectivity, evidence, and responsiveness to the study charge. The review comments and draft manuscript remain confidential to protect the integrity of the deliberative process.

We thank these individuals for their review of this report:

Duane Blackburn, The MITRE Corporation; **Frank Calzonetti**, University of Toledo; **Jim Cooney**, Colorado State University (ret.); **Arthur Ellis**, Elsevier; **Delores Etter**, Southern Methodist University (ret.); **Thomas Gold**, University of California, Berkeley; **Allan Goodman**, Institute of International Education; **Sheena Chestnut Greitens**, University of Texas at Austin; **Bruce Held**, U.S. Department of Energy (ret.); **James Holloway**, University of New Mexico; **S. Jack Hu**, University of Georgia; **Susan Pertel Jain**, University of California, Los Angeles; **Brendan Mulvaney**, National Defense University; **Pamela Norris**, The George Washington University; **Bill Priestap**, Trenchcoat Advisors LLC; and **Hank Reichman**, California State University-East Bay (ret.).

Although the reviewers listed above provided many constructive comments and suggestions, they were not asked to endorse the conclusions or recommendations of this report nor did they see the final draft before its release. The review of this report was overseen by **Julia Phillips**, Sandia National Laboratories (ret.) and **Jared Cohon**, Carnegie Mellon University (ret.). They were responsible for making certain that an independent examination of this report was carried out in accordance with the standards of the National Academies and that all review comments were carefully considered. Responsibility for the final content rests entirely with the authoring committee and the National Academies.

Acknowledgments

The committee acknowledges the U.S. Department of Defense for its support of this study.

ACKNOWLEDGMENT OF PRESENTERS

The committee gratefully acknowledges the contributions of the following individuals during open, public sessions held in support of the Confucius Institutes at U.S. Institutions of Higher Education consensus study:

April 8, 2022
- Bindu Nair, Director of Basic Research, Office of the Under Secretary of Defense for Research and Engineering, U.S. Department of Defense

April 29, 2022
- Joan Brzezinski, Executive Director, China Center, University of Minnesota
- Randy Kluver, Dean, School of Global Studies and Partnerships, and Professor, School of Media and Strategic Communication, Oklahoma State University

June 22, 2022
- Naima Green-Riley, Assistant Professor, Department of Politics and School of Public and International Affairs, Princeton University
- Denis Simon, Senior Adviser to the President for China Affairs, Duke University

- Sarah Spreitzer, Assistant Vice President and Chief of Staff, Government Relations, American Council on Education
- Richard Meserve, President Emeritus, Carnegie Institution for Science; Senior of Counsel, Covington & Burling LLP; Co-chair, National Science, Technology, and Security Roundtable, National Academies of Sciences, Engineering, and Medicine

July 20, 2022
- Jeffrey Lehman, Vice Chancellor and Professor of Law, New York University Shanghai
- Arun Seraphin, Deputy Director, Emerging Technologies Institute, National Defense Industrial Association
- Kevin Gamache, Associate Vice Chancellor and Chief Research Security Officer, Office of Research, The Texas A&M University System
- Emily Weinstein, Research Fellow, Center for Security and Emerging Technology, Georgetown University

August 16–17, 2022
- Nelson Dong, Partner, Dorsey & Whitney LLP
- Eli Friedman, Associate Professor and Chair of International and Comparative Labor, ILR School, Cornell University
- Peidong Sun, Associate Professor of History and Distinguished Associate Professor of Arts & Sciences in China and Asia-Pacific Studies, Department of History, Cornell University
- Yaqiu Wang, Senior China Researcher, Human Rights Watch

Contents

PREFACE   xiii

SUMMARY   1

1 **INTRODUCTION**   7
Growing Concerns, 10
Closures, 12
Report Purpose, Charge, and Approach, 13
Report Structure, 16

2 **CHARACTERISTICS AND FEATURES OF CONFUCIUS INSTITUTES**   17
Definition, 17
Current Landscape, 18
Attributes, 22
Contracts, 23
Relationship of Confucius Institutes with U.S. Campuses, 24
Relationship of Paired Chinese Institutions of Higher Education with U.S. Campuses, 24

3 **BENEFITS AND RISKS POSED BY CONFUCIUS INSTITUTES TO ACADEMIC INSTITUTIONS**   27
Benefits to U.S. Institutions of Higher Education Hosting a Confucius Institute, 27
Risks to U.S. Institutions of Higher Education Hosting a Confucius Institute, 29

4	**RISKS POSED BY CONFUCIUS INSTITUTES TO DOD-FUNDED RESEARCH**	**35**

Relationship Between Confucius Institutes and DOD-Funded Research, 35
Risks Posed by Confucius Institutes to DOD-Funded Research, 36

5	**FINDINGS**	**39**

Findings Regarding Background and Context, 39
Findings Regarding the Effect of CIs on Academic Freedom and University Governance, 40
Findings Regarding the Effect of CIs on DOD-Funded Research, 41

6	**RECOMMENDED CONDITIONS FOR GRANTING OF A WAIVER**	**43**
	REFERENCES	**49**
	APPENDIXES	
A	COMMITTEE BIOGRAPHICAL INFORMATION	57
B	LISTING OF OPEN, CLOSING, AND PAUSED U.S. CONFUCIUS INSTITUTES	65
C	OVERVIEW OF DOD-SPONSORED FUNDAMENTAL RESEARCH	67
D	CLOSURE REASONS FOR U.S. CIs USING NATIONAL ASSOCIATION OF SCHOLARS DATA	69
E	INFORMATION-GATHERING SESSIONS WITH U.S. COLLEGES AND UNIVERSITIES THAT ARE CURRENT OR FORMER HOSTS OF CONFUCIUS INSTITUTES	83

Preface

China has emerged as a global power, and its influence on the technological, industrial, and cultural landscape has grown enormously over the past 30 years. As leaders of U.S. institutions of higher education, we have long recognized the importance of understanding and engaging with China. We believe it is critical to develop U.S. citizens who understand Chinese language and culture and have deep expertise on China so that we can engage this emerging world power in a deep, nuanced, and clear-eyed way. Confucius Institutes (CIs), Chinese government-funded language and culture centers, have offered U.S. universities and their surrounding communities one pathway for building capacity in Chinese language and culture by expanding cocurricular programming and growing research relationships with Chinese partner universities.

And yet, the establishment of CIs on U.S. campuses comes with potential threats to cherished ideals and research security. We believe that the success of U.S. higher education rests crucially on protecting academic freedom as well as freedom of expression and dissent. Our students and faculty must be free to pursue any direction of intellectual inquiry they wish, and our campuses must be places where those of differing views engage in civil and respectful dialogue. It is this difference in values coupled with the strategic competition between the United States and China that gave rise to the concerns underlying this report.

We recognize the national security community's legitimate concerns regarding the presence of foreign-funded programs on campus, including the potential for espionage and intellectual property theft. We agree that research conducted on our campuses sponsored by the Department of Defense (DOD), the U.S. government, industry, and other organizations needs to be secured appropriately in order to protect innovations and mitigate risks, while maintaining the openness and international collaborations that allow America's research enterprise to flourish. Furthermore, we must be careful to distinguish the actions of the Chinese state from those of individual Chinese citizens, and especially faculty and students of Chinese origin on U.S. campuses, who have been invaluable contributors to scientific, technological, and economic progress. We acknowledge that the lack of distinction between the Chinese state and Chinese people can result in harmful and racist actions.

Responding to a request from Congress, the National Academies of Sciences, Engineering, and Medicine (the National Academies) formed an ad hoc committee to develop waiver criteria to potentially permit the continued presence of CIs on U.S. university campuses that also receive DOD funding. The committee's Statement of Task was developed with DOD in response to Section 1062 of the Fiscal Year 2021 National Defense Authorization Act (NDAA). The committee's charge for this report was to develop a set of waiver criteria that DOD could use to create a waiver process for use on the first day of Fiscal Year 2024 (October 1, 2023). Receiving a waiver would allow a U.S. institution of higher education hosting a CI and receiving DOD funding to continue to do both, given that appropriate safeguards are in place. The committee's focus has been to understand the attributes of CIs that create risks; to formulate waiver criteria that determine whether appropriate steps have been taken to mitigate those risks; and to ensure that these waiver criteria are straightforward to implement and are appropriate for an open campus. The committee sought only to formulate the criteria; DOD is expected to create and implement a workable waiver process. A second report, expected in June 2023, will address foreign-funded programs on campus and international partnerships more broadly.

The committee finds that CIs are one aspect of a long-term coordinated plan by the Chinese Communist Party to influence global perceptions of China and that the structure of a typical CI makes it a potential vehicle for direct oversight and intervention by the Chinese government. Rather than addressing CIs separately, DOD should adopt an integrated approach to addressing broader security concerns on U.S. campuses. Waiver criteria recommended by the committee include a demonstration by U.S. host institutions that they fully comply with all applicable DOD requirements for information, data, and research security; that they possess full managerial and fiduciary control of the CI; and that the CI adheres to key values of shared governance, openness, and academic freedom espoused by U.S. universities. Most importantly, if the host institution has demonstrated compliance with DOD's waiver criteria and waiver processes and if DOD is not aware of any relevant adverse information, a waiver should be granted in a reasonable time frame.

We would like to conclude with a note of deep appreciation for the hard work by the committee members who generously volunteered their time and expertise; the many experts who shared their knowledge and deep experience with our committee; the consultants at the American Institutes for Research who helped infuse our discussions with on-the-ground experiences of CI personnel; consultant writer Joe Alper, who helped shape our discussions into a coherent and readable form; and finally, the support provided by the staff at the National Academies, led by the extraordinarily able and efficient Sarah Rovito. We hope the diversity of opinions and experiences of this group have helped create a consensus report that will not only protect national security and academic freedoms but also enable U.S. universities to continue to be world leaders in higher education.

Philip J. Hanlon, *Chair*
Jayathi Y. Murthy, *Vice Chair*
Committee on Confucius Institutes at U.S. Institutions of Higher Education

Summary

More than 100 U.S. institutions of higher education hosted Confucius Institutes (CIs)—Chinese government-funded language and culture centers—on campus during the late 2000s and 2010s. While CIs provided a source of funding and other resources that enabled U.S. colleges and universities to build capacity, offer supplemental programming, and engage with the local community, CIs presented an added, legitimate source of risk to host institutions with respect to academic freedom, freedom of expression, and national security.[1]

By 2017, deteriorating U.S.-China relations led some U.S. colleges and universities to reconsider the value of having a CI on campus. This resulted in part from concerns by U.S. policy makers that Chinese entities on U.S. campuses might attempt to stifle criticism of the Chinese government; that China's increasing ability to translate its wealth into influence on American administrations, curricula, and public programs might have impacts counter to core academic principles; and that CIs could present a vulnerability to and conduit for espionage and intellectual property theft.

The committee is not aware of any evidence at the unclassified level that CIs were ever associated with espionage or intellectual property theft. While incidents affecting academic freedom, freedom of expression, and shared governance did take place, the most egregious of these happened at CIs outside of the United States. Sustained interest by Congress and political pressure led numerous

---

[1] The committee wants to state upfront that when referring to China, this report is referring to the People's Republic of China, the State, which is controlled by the Chinese Communist Party (CCP), and not to its people, many of whom are invaluable contributors to the global scientific enterprise (Albert et al., 2021). The CCP has more than 96 million members as of 2021, while China's total population is more than 1.4 billion people (Rui, 2022; World Bank, 2022).

U.S.-based CIs to close, especially following the passage of the Fiscal Year (FY) 2019 National Defense Authorization Act (NDAA), which contained a provision that ultimately barred institutions receiving Department of Defense (DOD) critical language flagship funding in Chinese from hosting a CI. While this provision allowed for a waiver process—and several affected colleges and universities applied for waivers in 2018 and 2019—DOD did not issue any waivers. Today, seven CIs remain on U.S. university and college campuses, most of which are at under-resourced institutions.

The committee's Statement of Task was developed with DOD in response to the FY 2021 NDAA, which contained a provision that barred institutions hosting a CI from receiving *any* DOD funding. However, in this provision, Congress allowed DOD to consult with the National Academies of Sciences, Engineering, and Medicine in considering waivers that could potentially provide a pathway for an institution hosting a CI to continue receiving agency funding. The Statement of Task directs the committee to deliver a first report after 12 months that recommends conditions that should be in place for DOD to consider granting a waiver to an institution of higher education, and a second report after 18 months that explores foreign-funded partnerships on U.S. campuses more broadly.

This report addresses the first charge and presents a set of findings and recommendations focused on waiver criteria that DOD can use to delineate a clear and transparent waiver process in advance of FY 2024. The committee's recommendations, developed after receiving input from a variety of key individuals and organizations during open committee meetings, are as follows:

Recommendation 1.1: Granting Waivers
In the absence of any applicable adverse information that cannot be addressed or mitigated through the criteria below or other means, DOD should grant a waiver if an applying institution of higher education meets the stated waiver criteria.
The committee recognizes that there may be classified reasons why a waiver might not be awarded to an individual institution of higher education. However, in the absence of DOD being aware of adverse information that cannot be addressed or mitigated, and in instances where all other criteria are satisfied, DOD should grant a waiver.

Recommendation 1.2: Communicating about Waivers
If DOD does not grant any waivers, or decides not to grant a waiver to a specific institution of higher education when other waivers are awarded, it should specify the reason(s) for denial to the extent possible at the unclassified level.
This will potentially allow an institution denied a waiver to understand underlying concerns and to address security risks on campus that they may not be aware of. Colleges and universities also may want to consider having access to or

retaining a cleared individual in order to have a fuller understanding of security-related issues.

Recommendation 1.3: Establishing the Waiver Application Process
Outside input is critical to ensure that the waiver application process is free from undue administrative and regulatory burden. In addition to U.S. government input, DOD should solicit external input from key organizations, including industry, higher education associations, and universities.
DOD should specify whether this is a one-time, permanent waiver or provide additional information regarding the duration of the waiver and process for subsequent application, evaluation, and renewal.

Recommendation 2: Waiver Criteria
  a. U.S. host institutions should demonstrate that the CI is a formally established Center or Institute at the institution, thereby subjecting the CI to all policies and procedures prescribed in faculty, staff, and student codes, as well as in shared governance documents that ensure that similar units within the university support the key values of American academic institutions, including academic freedom and openness and respectful behavior toward other host institution academic units. If a U.S. host institution is not structured in a way that allows for formal Centers or Institutes, it should develop a structure for oversight and include the details for that structure in the documents governing a CI.
      i. U.S. host institutions should provide documentation of relevant sections on Centers and Institutes in their institutional handbook or policy manuals.
      ii. U.S. host institutions should provide documentation of an established, regular review process for Centers and Institutes that would ultimately include the CI and would specify the period/frequency of review. This would include external reviews and advisory councils, as required by the institution.
      iii. U.S. host institutions should provide documentation of public statements on academic freedom and freedom of expression as codified in university policy.
      iv. U.S. host institutions should provide documentation of posted CI bylaws and governance documents, including operational and administrative policies and practices.
  b. U.S. host institutions should demonstrate that they meet and comply with all applicable DOD requirements for information, data, physical, and research security.
      i. U.S. host institutions conducting $50 million or more of federally funded research per year should demonstrate compliance with

National Security Presidential Memorandum – 33 (NSPM-33) or subsequent versions of this document. These institutions can satisfy this criterion by demonstrating that they have established and operate a research security program that includes elements of cybersecurity, foreign travel security, insider threat awareness and identification, and export control training.

    ii. U.S. host institutions conducting less than $50 million of federally funded research per year should demonstrate the presence and implementation of adequate research security measures on campus. These institutions can satisfy this criterion by providing research security-related documents, including elements of cybersecurity, foreign travel security, insider threat awareness and identification, and export control training.

    iii. U.S. host institutions should demonstrate that they have appropriate safeguards in place to ensure that CI faculty and visitors who are not university employees have limited or guest access to university computer networks and cannot access networks that store research results and communications. Institutions can satisfy this criterion by providing the university's cybersecurity and visitor network access policy and demonstrating that CI faculty and visitors have limited or guest access.

c. U.S. host institutions should demonstrate that they possess full managerial control of CI curriculum, instructors, textbooks and teaching materials, programmatic decisions, and research grants.

    i. U.S. host institutions should demonstrate that CI employees and affiliates are formally associated with the host institution and subject to human resources policies and procedures. Institutions can satisfy this criterion by providing documentation, such as an employment contract or agreement, that the director of the CI is employed by the university with a reporting line to the host institution's chief academic officer or their designee, and by providing public-facing personnel rosters that clearly state whether the host institution classifies CI-affiliated personnel from the Chinese partner institution as either host institution employees or as visiting scholars. CIs should hire their employees and affiliates in accordance with the host institution's human resources policies and procedures and subject to corresponding campus policies.

    ii. U.S. host institutions should demonstrate that CI curricula, including syllabi, textbooks, and teaching materials, are approved through faculty governance review.

    iii. U.S. host institutions should demonstrate oversight over CI-supported research grants. In addition, universities should submit

conflict of interest, conflict of commitment, and export control protocols that demonstrate university control.
d. U.S. host institutions should ensure that no contract or other written agreement pertaining to creating or operating the CI calls for the application of foreign law to any aspect of the CI's operation at any U.S. campus of the host institution.
e. U.S. host institutions should demonstrate appropriate fiduciary and financial oversight of the CI. Host institutions can satisfy this criterion by providing the following:
   i. A publicly available annual budget for the CI, including sources of revenue and expenses.
   ii. A copy of the agreement between the host institution and Hanban[2] to host a CI (a hard or electronic copy of the original document, including the original version of the agreement in Chinese). The host institution should manage the agreement through its sponsored program process.
   iii. A copy of the Memorandum of Understanding and contract between the U.S. host institution and Chinese partner institution, if applicable.
   iv. A copy of the policy that any financial contribution from foreign or domestic sources supporting the CI must be treated as a sponsored contract, not a gift, with a deliverable (programming, education, etc.).

The committee is optimistic that these waiver criteria will be useful to DOD as the agency formulates a waiver process in the coming months. In the meantime, the committee will continue its work and broaden its exploration to include other foreign-funded partnerships on U.S. campuses and to identify implementable practices and principles regarding appropriate operations for academic institutions in accordance with its Statement of Task. The committee will present additional findings and recommendations in a second report to be released in June 2023.

---

[2] "Hanban" is the colloquial term for the Chinese International Education Foundation, or CIEF, now known as the Ministry of Education Center for Language Education and Cooperation, or CLEC. This is the Chinese government agency affiliated with China's Ministry of Education that promoted, managed, and funded CIs on foreign campuses.

1

Introduction

Confucius Institutes (CIs) are Chinese government-funded centers whose purpose is to improve the worldwide opinion of China by offering classes in Mandarin Chinese and highlighting positive aspects of Chinese culture. CIs are modeled on similar cultural programs, such as those of the UK's British Council, Germany's Goethe Institutes, and France's Alliance Française (CRS, 2022). While these European public diplomacy initiatives serve as models, one significant difference is that these programs do not bear the stamp of any one British, German, or French political party. The Chinese Communist Party (CCP), via the Chinese International Education Foundation (formerly Hanban), designs and oversees the CI initiative (Sahlins, 2014). The CCP views education, even in China, more as a means for economic development and power projection than, as in the U.S.'s conception, for the creation and transmission of knowledge and skills that improve individual well-being and benefit society as a whole (Brady, 2017; Mosher, 2012).[1]

China's first CI opened in Seoul's Gangnam District in 2004, followed by its first U.S.-based CI later that year at the University of Maryland, College Park. Eventually, the program grew to enroll more than 9 million students at 548 institutions in 146 countries,[2] including 121 institutions in the United States.

---

[1] The committee wants to state upfront that when referring to China, this report is referring to the People's Republic of China, the State, which is controlled by the Chinese Communist Party (CCP), and not to its people, many of whom are invaluable contributors to the global scientific enterprise (Albert et al., 2021). The CCP has more than 96 million members as of 2021, while China's total population is more than 1.4 billion people (Rui, 2022; World Bank, 2022).

[2] See https://www.heritage.org/homeland-security/commentary/confucius-institutes-Chinas-trojan-horse.

Most U.S.-based CIs were on college and university campuses, including those of state flagship institutions, private schools, and historically Black colleges and universities.

At their peak, a number of U.S.-based CIs did not offer credit-bearing courses to enrolled students; rather, they focused on education for part-time learners and offered K–12 programs through the related Confucius Classrooms program. Confucius Classrooms provide curriculum guides, textbooks, technology, and teachers for U.S. children, including children in distressed school districts who could not otherwise study Chinese. Given its focus on CIs at institutes of higher education, this report does not explore Confucius Classrooms in further detail; however, the committee acknowledges the program's existence, audience, and impacts within the context of discussing CIs (Green-Riley, 2020).

At the outset, a Chinese government agency called Hanban, which was affiliated with China's Ministry of Education and led by Vice Minister Xu Lin, promoted and managed CIs. The original Hanban model called for China to provide seed money for CIs that foreign host institutions would match. Over time, Chinese financial support was supposed to be phased out as CIs became self-sustaining, but this rarely happened. In the United States and most other countries, local directors hired by the host university as employees in faculty or staff roles managed the CIs (GAO, 2019). These individuals worked in consultation with Chinese instructors approved and dispatched by Chinese partner schools.

Most CI cultural programs have been described as apolitical by design, but they do advance a heavily curated view of China. They focus on Chinese visual and performing arts, aesthetics, uncontroversial if selective aspects of the distant past, and the beauty of China's landscapes. They do not offer in-depth courses on Confucianism or cultural criticism, nor do they engage in direct political indoctrination. The CI program's Chinese designers may know of American political and ideological concerns and may not want to threaten their primary language programs by defending, for example, China's claims in the South China Sea, its currency policies, or its human rights practices. Hanban's general approach is to build China's soft power by presenting China to foreign publics simply as an unthreatening developing nation with rich traditions (Ren, 2012).

Despite this programmatic caution, there were early signs that CCP practices pervaded China's global oversight of CIs. The most striking evidence of the CCP's hand was that many early Memorandums of Understanding (MOUs) establishing CI partnerships between Chinese and U.S. universities were beholden to confidentiality clauses and not made public (GAO, 2019). This practice is uncommon for U.S. university MOUs with foreign partners. Such a lack of transparency increased the suspicions of Chinese language and Chinese studies faculty at many U.S. colleges and universities, who often opposed the creation of CIs as duplicative and as undermining their own work. In some examples, university presidents, and not faculty, led the charge to host a CI (GAO, 2019). Other concerns were that CIs would operate without faculty input and might saddle campuses with compromises that would not hold up under scrutiny. Because CIs were answerable to the Chinese government, at least in part, faculty perceived them to be counter to the principle of academic self-governance (Sahlins, 2014).

The actions of Vice Minister Xu were also cause for concern. In 2014, the European Association for China Studies complained that Xu, while attending an association conference in Portugal, had decided that some presenters' abstracts "were contrary to Chinese regulations" and had objected to the conference program's characterization of Taiwan's Chiang Ching-kuo Foundation which, like Hanban, sponsored the event. Without consultation, Xu instructed her staff to confiscate the materials and later returned them with the offending pages removed.[3]

Other incidents also raised concerns in the academic community. When the Chinese government offered to fund a new professorship at Stanford University, whose CI had opened in 2013, a Hanban official "expressed concern that (an) endowed professor might discuss 'politically sensitive things, such as Tibet'" (Golden, 2015; Redden, 2012). Stanford refused to establish a CI under such terms, and Hanban agreed to move forward in funding the CI without this restriction in place. The university pushed back a second time when CI Chinese partner institution Peking University insisted that Stanford's CI focus on social sciences (Knowles and Jin, 2019). Furthermore, North Carolina State University, whose CI opened in 2006 and closed in 2018, reportedly turned down a visit by the Dalai Lama at the suggestion of its Chinese CI partner institution (Golden, 2015).

Canada's McMaster University closed its CI after a former instructor filed a discrimination complaint alleging that she was prohibited from participating in the spiritual practice Falun Gong by language in her contract with Hanban (Redden, 2017). McMaster University officials "sought to find a way to get around ... discriminatory hiring practices but when they determined no solution was possible, they decided to close their institute." This led at least one U.S. CI to add a nondiscrimination clause to its contract with Hanban, as the University of Iowa updated its contract language to state that its CI would "hire all employees internally" and not allow Hanban to "interfere with hiring practices" (Riley, 2013). Journalist Bethany Allen-Ebrahimian wrote about her experience of having Taiwan edited out of her biography by the co-director of Savannah State University's CI when receiving an award from the institution's Department of Journalism and Mass Communications (Allen-Ebrahimian, 2018). Such incidents sparked opposition to CIs from the National Association of Scholars, the American Association of University Professors, and the University of Chicago, among other groups (AAUP, 2014; Peterson, 2017; Schmidt, 2014).

The number of CIs continued to expand from 2004 to 2019 as universities sought to catalyze lucrative relationships and collaborations with Beijing and as CIs, and as Confucius Classrooms in particular, met a growing demand for Chinese language instruction. The expansion of CIs was not an isolated phenomenon

---

[3] See https://www.wsj.com/articles/beijings-propaganda-lessons-1407430440.

and occurred simultaneously with other changes, the combination of which raised alarms. The CI movement corresponded with

- ongoing increases in the costs of tuition, room, and board on U.S. campuses;
- ongoing decreases in the number of domestic students enrolling at U.S. campuses (NCES, 2022; NSCRC, 2022);
- an explosion in U.S. enrollments and tuition payments by Chinese undergraduates following the financial crisis of 2008 (Feldgoise and Zwetsloot, 2022; Ma, 2020; OpenDoors, 2021);
- a rapid expansion in Chinese research and development expenditures along with increasing rates of Chinese intellectual property theft (Financial Times, 2022; NSB, 2022);
- a growing interest in academic development offices to build relationships with and secure gifts from Chinese corporations, Chinese individuals, and Chinese philanthropic foundations; and
- strong interest by the U.S. scientific community to collaborate with Chinese researchers, many of whom are leaders in their disciplines and have access to generous Chinese government funding and research infrastructure.

CIs were partnerships; China did not impose them on institutions in the United States, and universities entered them voluntarily. While many higher education and research partnerships between the United States and China, including and beyond CIs, continue, such collaborations have become increasingly complex. These collaborations are emblematic of an era in which the United States and China became more deeply integrated with each other academically, commercially, and geostrategically. Distinctions between scholarly, economic, and national interests were blurred and easily confused. The committee firmly believes that this era has ended given the rise of a new geostrategic environment in the Indo-Pacific region.

GROWING CONCERNS

By 2017, deteriorating U.S.-China relations had begun to fuel broader suspicion by the U.S. government of close financial and research ties between U.S. universities and their Chinese counterparts. Furthermore, a misalignment in values between the United States and China became evident. Concerns that Chinese entities on U.S. campuses might attempt to stifle criticism of the Chinese government; augment China's growing ability to translate its wealth into influence on U.S. administrations, curricula, and public programs; and introduce potential mechanisms for espionage and intellectual property theft posed by the openness of American universities caused U.S. policy makers to reevaluate academic ties with China. Meanwhile, administrators at U.S. institutions of higher education were left to weigh the benefits of keeping their CIs open with the costs of increased congressional scrutiny.[4]

U.S. legislators' sustained attention to CIs and China's reach into the U.S. educational system proved more consequential than the earlier complaints of U.S. academic sinologists. Congressional alarm over the alleged influence of CIs coincided with, and was exacerbated by, lawmakers' concerns over the strategic implications of a number of Chinese programs, policies, and activities that included the following:

- The *Indigenous Innovation* agenda, launched in 2006,[5] that called on Chinese universities and scientific institutes to develop technology to build China's wealth and power.
- *Made in China 2025*, a program announced in 2015 that listed specific industries and technologies that China intended to dominate and lead.
- *Military-Civil Fusion* policies, dating to the 1990s, that called on all Chinese institutions to share relevant technologies with the People's Liberation Army.
- *National Intelligence Laws*[6] requiring Chinese entities, including universities, to give the Chinese government any information or data it requested.
- The *Thousand Talents* program, along with other state-sponsored talent recruitment schemes, that sought to attract foreign science, technology, engineering, and mathematics experts, particularly those of Chinese origin, to China to contribute to Chinese President Xi Jinping's national rejuvenation agenda (Weinstein, 2022). Scholars in the *Thousand Talents* program were a focus of the U.S. Department of Justice's *China Initiative*, launched by the Trump administration, that targeted Chinese attempts to steal technology from U.S. universities and corporations (DOJ, 2021).
- The *Huawei Corporation's* dominance of 5G technology and possible network security concerns, along with growing awareness of its funding of science, technology, engineering, and mathematics programs on U.S. campuses (Flaherty, 2019; Friis and Lysne, 2021; Gibney, 2019).

In Washington, D.C., CIs—like the activities listed above—became another example of a Chinese activity of concern that either took place within U.S. borders or aimed to harm U.S. interests.

---

[4] See, for example, https://www.rubio.senate.gov/public/index.cfm/2018/2/rubio-warns-of-beijings-growing-influence-urges-florida-schools-to-terminate-confucius-institute-agreements.

[5] See https://www.Chinalawinsight.com/2010/09/articles/intellectual-property/Chinas-indigenous-innovation-policy-and-its-effect-on-foreign-intellectual-property-rights-holders/.

[6] See https://www.lawfareblog.com/beijings-new-national-intelligence-law-defense-offense.

CLOSURES

Between 2016 and 2020, several laws, U.S. Department of Education investigations, and other federal actions whittled away at the standing and viability of CIs and targeted their operations. The August 2018 signing into law of the John S. McCain National Defense Authorization Act (NDAA) for Fiscal Year 2019 created new requirements for institutions of higher education to host CIs and led to numerous U.S.-based CIs closing. The NDAA prevented the Department of Defense (DOD), which funds key critical language programs at universities across the country, from supporting Chinese language study at any school that housed a CI unless DOD issued a waiver.[7] To date, DOD has not issued a single waiver. In addition, the U.S. Department of State designated the Confucius Institute U.S. Center, the headquarters of the American CI network, as a foreign mission of the People's Republic of China on August 13, 2020.[8]

Although a bipartisan committee had found no evidence of espionage, intellectual property theft, or any other illegal activity being conducted by CIs,[9] U.S. campuses faced a Hobson's choice.[10] U.S. universities that took Chinese language instruction seriously had to decide whether to take money from the U.S. government or the Chinese government. Other institutions without federally supported language programs also received significant pressure to close their CIs. The 2021 NDAA increased the penalty for universities that hosted CIs, as it prohibited them from receiving DOD funding *for any purpose* if their CIs remained open, unless DOD issued a waiver.

As of December 2022, the number of CIs operating at U.S. institutions of higher education has dropped from 121 to 7, one of which will close in 2023.[11] While the 2019 NDAA created difficult choices for the affected institutions, anecdotal evidence indicates that some used the new requirements to shut down programs that had become a financial drain or had otherwise underperformed (Zialcita, 2021). Worsening U.S.-China relations, the reduction in the number of Chinese students studying in the United States,[12] and the drop in U.S. interest in Chinese language study[13] may have undermined the original justification for CIs. However, the declining number of CIs does not necessarily indicate a declining

---

[7] John S. McCain National Defense Authorization Act for Fiscal Year 2019, Public Law 232, 115th Congress 2nd Session (August 13, 2018).

[8] See https://2017-2021.state.gov/confucius-institute-u-s-center-designation-as-a-foreign-mission/index.html.

[9] See https://www.hsgac.senate.gov/subcommittees/investigations/media/senators-portman-and-carper-unveil-bipartisan-report-on-confucius-institutes-at-us-universities_k-12-classrooms.

[10] Merriam-Webster defines "Hobson's choice" as "an apparently free choice when there is no real alternative."

[11] See Appendix B, "Listing of Open, Closing, and Paused U.S. Confucius Institutes."

[12] See https://www.wsj.com/articles/Chinese-student-visas-to-u-s-tumble-from-prepandemic-levels-11660210202?page=1.

[13] See https://www.insidehighered.com/news/2019/01/09/colleges-move-close-chinese-government-funded-confucius-institutes-amid-increasing.

interest in academic partnerships for language studies. A recent study by the National Association of Scholars indicates that some U.S. colleges and universities, including institutions that formerly hosted CIs, may be conducting similar activities under different auspices, with direct or indirect financing from China.[14]

REPORT PURPOSE, CHARGE, AND APPROACH

The criteria DOD uses to determine whether the arrangements between U.S. institutions of higher education and their CIs warrant granting waivers is a pressing matter for institutions of higher education; the U.S. government; and if the most dire warnings of CI critics have merit, the American people. This committee of the National Academies of Sciences, Engineering, and Medicine (the National Academies) was asked by DOD, in response to Congress, to recommend criteria that colleges and universities should meet if the agency is to consider granting a waiver to the prohibition of DOD research support to an academic institution hosting a CI in a first report and to address additional issues concerning foreign-funded institutes on U.S. campuses and international partnerships in a second report (see Statement of Task in Box 1-1).

To carry out this charge, the National Academies formed a committee of leaders and scholars that included higher education administrators and researchers, science and technology policy experts, foreign language and China experts, international programs experts, and national security experts. Members of the committee unanimously support international partnerships, the core values of academic freedom and academic self-governance, and the openness and global character of the scientific and research enterprise. The committee also believes strongly that the study of critical languages is essential to national security.

Given its interpretation of the Statement of Task writ large, the committee addressed the following elements in this first report in service of providing a set of findings and recommendations, including waiver criteria to DOD:

- Identifying existing CIs and conducting case studies of existing and recently closed or soon-to-be-closed CIs to understand attributes of the relationship between CIs and their respective U.S. host institutions of higher education.
- Exploring why and how closure or renewal decisions were made.
- Examining policies and processes that open CIs have in place to protect against undue foreign influence that could adversely impact the academic education and research environment.

The committee will address the remaining components of the Statement of Task in a second report, to be released no later than June 30, 2023:

---

[14] See https://www.nas.org/reports/after-confucius-institutes/full-report.

> **BOX 1-1**
> **STATEMENT OF TASK**
>
> In light of recent restrictions placed on DOD that would prohibit support for research conducted at academic institutions that host Confucius Institutes, an ad hoc committee of the National Academies of Sciences, Engineering, and Medicine will conduct a study of Confucius Institutes in the United States and the potential risks these Institutes may pose to academic institutions and by extension to DOD-funded academic research.
>
> Specifically, the committee will identify existing Confucius Institutes and conduct case studies of existing and recently closed or soon-to-be-closed Institutes, to understand the attributes of the relationship between Institutes and their U.S. host institutions of higher education. The study will include a consideration of such things as the characteristics of funding and reporting requirements, and foreign government involvement in staffing, curriculum, and other Institute activities. Further, the study will be conducted within the context of possible threats to the inherent values of U.S. academic institutions such as openness and free expression of thought in instruction and scientific research, and the ethos of scientific research such as research integrity, reciprocity, and the free flow of talent and ideas.
>
> In cases where an Institute has been closed or will not be renewed, the committee will explore why these decisions were made, whenever possible. In cases where Institutes will continue, the committee will examine the policies and processes the academic institutions have in place to guard against any untoward foreign government influence that might distort the academic education and research environment.

- Gathering information on other foreign-funded institutes at U.S. institutions of higher education and describing characteristics and features of such institutes.
- Determining characteristics and features of foreign-funded institutes at U.S. institutions of higher education that could be flags for institutions to engage in further deliberation and vetting prior to entering into a partnership.
- Identifying implementable practices for U.S. institutions of higher education to ensure appropriate operations.
- Continuing exploration of what role the sensitivity of the research conducted on campus should play in determining which foreign-funded partnerships are appropriate.

An extensive literature review and study of relevant legislation, think tank reports, and recommendations issued by higher education associations, including

INTRODUCTION

> While the focus of this study is on Confucius Institutes, the committee will also seek information on other foreign-funded institutes at U.S. institutions, to the extent possible.
>
> Drawing on its analysis of Confucius Institutes, the committee will describe some of the characteristics and features of foreign-funded programs at U.S. institutions that could serve as flags for academic institutions that would lead to further deliberation and vetting to determine whether or not to enter into a partnership. As the Confucius Institute program evolves, these characteristics and features will allow academic institutions and DOD to identify programs that may not meet academic criteria for openness and independence in education and research.
>
> Additionally, the committee will identify best practices and principles regarding appropriate operations for U.S. academic institutions that DOD could use as it makes its determinations regarding whether to grant or deny waivers to the prohibition of support for research conducted at academic institutions that host Confucius Institutes.
>
> Finally, the committee will consider what, if any, role the sensitivity of the research conducted on campus (unclassified, controlled, classified) should play in determining what foreign-funded international partnerships like Confucius Institutes are appropriate.
>
> The committee will deliver an interim report at 12 months on what conditions it recommends be in place for DOD to consider granting a waiver of its prohibition against research support for an academic institution hosting a Confucius Institute, along with any other preliminary findings the committee may have.
>
> Following its 18-month analysis, the committee will issue a consensus report with findings and recommendations to DOD on the issues described above.

the American Council on Education, the Association of American Universities, the Association of Public and Land-grant Universities, and the Council on Governmental Relations and by other organizations, including the Hoover Institution and the JASON independent science advisory group, have informed this study. In addition to considering reports and recommendations from a range of issue experts, the committee commissioned the American Institutes for Research to interview U.S. personnel from institutions currently and formerly hosting CIs to obtain deeper insights into their operations, viewpoints, and concerns.

The National Academies and DOD agreed to conduct this study at the unclassified level to ensure maximum transparency and accessibility and to engender trust between institutions of higher education and the agency. Both parties felt that this was necessary for communicating the justification for best practices and protections and for the development and implementation of a fair and effective waiver process—a waiver process that would enable institutions of higher education to understand what is being asked of them about academic freedom and national security risks and threats and to take appropriate mitigating actions in service of being considered for a waiver.

However, this decision led to the committee working with a deficit of information throughout the information-gathering phase of the study, as it was unable to consider classified data, evidence, and threats. Multiple speakers from federal government agencies, including the Federal Bureau of Investigation, were invited but declined to speak with the committee in support of the study. While the Department of State declined to speak with the committee, the agency did furnish written answers to a set of questions developed by the committee. The lack of engagement with the law enforcement and intelligence communities hindered the committee's exploration and understanding of security issues presented by CIs to U.S. institutions of higher education. The committee notes that this does *not* mean that adverse information or legitimate concerns regarding the presence of CIs and other foreign-funded entities on U.S. campuses does not exist. Indeed, the committee sought to conduct a rigorous exploration to the extent possible while working and developing findings, recommendations, and waiver criteria with available unclassified information.

REPORT STRUCTURE

The remainder of this report addresses the committee's activities, findings, and recommendations. Chapter 2 describes the current landscape, characteristics, and features of CIs. Chapters 3 and 4 discuss the benefits and risks posed by CIs to academic institutions and DOD-funded research, respectively. Chapter 5 presents the committee's findings, and Chapter 6 provides the committee's recommended conditions for granting a waiver.

The committee, through the information gathering conducted in support of this report, finds it is possible to implement measures on campus to mitigate—but not to fully eliminate—the risks associated with the presence of a CI and protect academic freedom, freedom of expression and dissent, and national security. Therefore, the committee developed waiver criteria it believes will protect U.S. academic values, research integrity, and security while allowing for expanded Chinese language study, which it believes is in the national interest. DOD can draw upon the recommended waiver criteria to create a process that will enable it to discern whether an institution of higher education has taken appropriate precautionary measures to receive a waiver.

It is the committee's hope that this study will contribute to safeguarding U.S. higher education and the U.S. innovation system while promoting the vigilant openness that has made U.S. colleges and universities among the finest in the world.

2

Characteristics and Features of Confucius Institutes

This chapter presents an overview of the characteristics and features of Confucius Institutes (CIs). As a result of its fact-finding activities, the committee acknowledges that the structure, management, and programming of CIs vary widely across institutions, and thus the committee could not develop an all-encompassing description of a CI's structure and how it functions on U.S. institution of higher education campuses. Rather, the committee's goal for this chapter is to provide a sense of how CIs operate in the United States.

DEFINITION

Section 1062 of the William M. (Mac) Thornberry National Defense Authorization Act (NDAA) for Fiscal Year (FY) 2021, which called for the National Academies of Sciences, Engineering, and Medicine to conduct this consensus study, defines the term "Confucius Institute" as a "cultural institute directly or indirectly funded by the Government of the People's Republic of China."[1] Later legislation expanded the definition of CI to a "cultural institute established as a partnership between a United States institution of higher education and a Chinese institution of higher education to promote and teach Chinese language and culture that is funded, directly or indirectly, by the Government of the People's Republic of China."[2]

---

[1] William M. (Mac) Thornberry National Defense Authorization Act for Fiscal Year 2021, Public Law No. 116-283, 116th Congress, 2nd Session (January 1, 2021), Section 1062 (U.S. Congress, 2021).

[2] H.R. 4346 – Supreme Court Security Funding Act of 2022, Public Law No. 117-167, 117th Congress (August 9, 2022) (U.S. Congress, 2022).

For this report, the CI definition encompasses CIs established on U.S. campuses as early as 2004 with funding from the Office of Chinese Language Council International, part of China's Ministry of Education. This funding organization was formerly known as Hanban, or CI Headquarters, and has since been renamed the Ministry of Education Center for Language Education and Cooperation (CLEC) (CRS, 2022; Peterson et al., 2022). CLEC's Chinese International Education Foundation currently oversees and funds existing CIs (Peterson et al., 2022).[3] The committee does not intend for this definition to capture Chinese language and culture partnerships funded through other nongovernmental funding channels, scientific and technological research partnerships and collaborations, or the Chinese Students and Scholars Association organization.

CURRENT LANDSCAPE

The committee identified 10 open CIs and 7 closing or paused CIs in the United States in March 2022 and is aware of 7 open CIs[4] in the United States as of December 2022.[5] Of the 7 open CIs, only 3 submitted data to the National Science Foundation's Higher Education Research and Development Survey (HERD) for FY 2020 (see Table 2-1). The committee notes that only 1 institution hosting a CI is above the $50 million federal funding threshold for compliance with National Security Presidential Memorandum—33, or NSPM-33. Given that all institutions performing $150,000 or more of research and development per year are required to participate in the National Science Foundation's HERD survey, it can be inferred that all but 1 of the 7 remaining institutions hosting CIs are below the $50 million federal funding threshold. Furthermore, the University of Utah and Alfred University are the only institutions hosting a CI and performing a sizable amount of DOD-funded research.

The committee discovered through the release of the National Association of Scholars' report *After Confucius Institutes: China's Enduring Influence on American Higher Education* in June 2022 that CIs at Bryant University, Medgar Evers College, and Presbyterian College had closed. The CI at St. Cloud State University closed in December 2021, and the CIs at Southern Utah University, the University of Akron, and the University of Toledo closed during summer 2022.[6] The CI at Stanford University closed on October 31, 2022.[7]

---

[3] This report refers to Hanban throughout, given that it was the funding organization's name when the CIs that the committee reviewed were established.

[4] This includes the University of Utah's CI, which is slated to close in June 2023.

[5] See Appendix B, "Listing of Open, Closing, and Paused U.S. Confucius Institutes."

[6] See https://www.uakron.edu/im/news/ua-to-close-confucius-institute. A reviewer of this Consensus Study Report provided further informational comment regarding the closing of the CI at the University of Toledo.

[7] Confirmation of the closing of the CI at Stanford University was provided to the study staff by email correspondence.

TABLE 2-1 Research and Development Expenditures for Remaining CIs for FY 2020

Institution	FY 2020 Total USG R&D Expenditures	FY 2020 DOD R&D Expenditures	FY 2020 NSF R&D Expenditures
Alfred University	$947,000	$597,000	$97,000
Pacific Lutheran University	-	-	-
San Diego Global Knowledge University	-	-	-
Troy University (Alabama)	$1,591,000	$0	$165,000
University of Utah	$314,023,000	$30,944,000	$34,587,000
Webster University (Missouri)	-	-	-
Wesleyan College (Georgia)	-	-	-

NOTE: NSF = National Science Foundation; R&D = research and development; USG = U.S. government.
SOURCE: National Science Foundation, HERD survey Table 25 (NSF, 2021).

While not all of these institutions shared the reasoning behind their CI closures, Bryant University President Ross Gittell stated in a 2021 letter to the community that the CI closure was the result of "changes that are taking place in China and regarding U.S.-China relations…" (Bryant University, 2021; NBC 10, 2021). Presbyterian College attributed its closure to staffing issues because of the COVID-19 pandemic (Peterson et al., 2022). Meanwhile, the University of Akron cited the restrictions on DOD funding found in Section 1062 of the FY 2021 NDAA, from which this report originated: "The University of Akron has developed a broad research partnership with the Department of Defense and with other funding agencies and organizations at different levels of the federal and state government. In order to meet the requirements of the NDAA, The University of Akron has decided to close its Confucius Institute…" (University of Akron News, 2021).[8] Valparaiso University, which closed its CI in March 2022, mentions the NDAA provision in its closure announcement as well (Padilla, 2021).

From the information-gathering activities the American Institutes for Research (AIR) conducted in support of this study,[9] the committee discovered that the provision in the FY 2019 NDAA served as the "deciding factor for some

---

[8] The University of Akron had more than $10 million dollars of federal research and development (R&D) expenditures in FY 2020, inclusive of $2,779,000 of DOD R&D expenditures (NSF HERD data for FY 2020, Table 25) (NSF, 2021).

[9] See Appendix E, "Information-Gathering Sessions with U.S. Colleges and Universities That Are Current or Former Hosts of Confucius Institutes."

institutions to close their CI[s]" (Kaleem et al., 2022).[10] One institution interviewed by AIR decided to close its CI "instead of applying for a waiver because they wanted to ensure that their federal funding was not jeopardized" (Kaleem et al., 2022). This is similar to the reasoning publicly cited by the University of Minnesota and the University of Rhode Island when announcing the closure of their respective CIs in February 2019 and the University of Maryland when announcing the closure of its CI in January 2020 (Bullard, 2020; Sabrowsky, 2019; Thennarasu, 2019).

The National Association of Scholars delves deeper into the reasons for closure of 104 U.S. CIs in its report *After Confucius Institutes: China's Enduring Influence on American Higher Education* (Peterson et al., 2022). The publicly stated reasons for closure collected by the National Association of Scholars were analyzed by National Academies staff (Appendix D). Institutional responses to U.S. public policy changes tended to be the most cited reason for closure, followed by a misalignment of the campus CI with institutional values or goals, and state or institutional budgetary reasons. The committee notes that further research is needed both to verify publicly stated reasons for closure and to understand how universities may have restructured or rebranded CI-like programs and partnerships going forward.

The 2018 audit that Tufts University conducted of its CI provides valuable insight into why or how closure and renewal decisions were made. The goals of the audit were to assess the benefits and concerns associated with the CI's operation, to recommend whether to renew the CI or not, and to recommend changes to the CI agreement(s) and governance if the CI was indeed to continue (Tufts, 2019). The audit committee sought to ensure that the evaluation was comprehensive and included the following:

- A review of CI-related documentation including draft and final contracts, workplans, annual reports, budgets, correspondence, and books and materials;
- A review of external literature, studies, and reports on CIs including many of the same documents utilized by the committee in service of this report;
- Interviews with CI staff and Tufts faculty and staff involved with negotiating CI agreements;
- Discussions with Tufts government relations advisors and Massachusetts congressional delegation staff; and
- Input collected from the community via open forums on their experience with and perspectives on the CI (Tufts, 2019).

---

[10] One respondent interviewed by AIR shared that the decision to close their institution's CI came directly from the university president, while another respondent stated that the decision was made by university leadership.

Tufts' audit found that the university engaged in "significant due diligence" prior to agreeing to establish a CI and in its negotiations with Hanban (Tufts, 2019). The audit also found that Tufts' CI-related agreements and practices aligned with external best practices for CIs; that the CI contributed to the teaching of Chinese language and culture at Tufts; that the CI had not exercised undue foreign influence on the university, faculty, staff, or students; and that the presence of a CI on campus raised reputational and ethical concerns. The audit committee noted that "the arguments both for renewal and non-renewal are strong" and ultimately decided to renew the CI for a 2-year period after adopting measures to fortify Tufts' governance of the CI (Tufts, 2019, 2021). However, Tufts announced in March 2021 that it would not renew the CI at the conclusion of the 2-year period in September 2021 (Tufts, 2021).

Of the CIs that remain open, the CI of the State of Washington (CIWA) at Pacific Lutheran University (PLU) posts its Certificate of Authorization from the Chinese International Education Foundation and Five Party Agreement (Memorandum of Understanding [MOU] between U.S. institutions and Chinese institutions) on its website.[11] In addition, CIWA's website notes that "all CIWA-supported programs are entirely designed and/or supervised by directors at PLU and SPS [Seattle Public Schools] in coordination with colleges or universities, K–12 educational institutions, community groups or individual partners across Washington state" (CIWA, 2022).[12] The CI at Wesleyan College similarly links to an open letter from the university president and an implementation agreement (MOU between Wesleyan College and Guangzhou University) on its website.[13] The former document emphasizes that "nothing in our contract 'leads to academic self-censorship of topics the Chinese government considers taboo,'" "Wesleyan College does not hesitate to discuss and teach about the topics that are considered taboo on many university campuses in China," and that the CI is of value to the institution and its surrounding community (Fowler, 2018). The CI at Troy University shares an email address for the public to use to obtain a copy of its annual report.[14] None of the other open CIs or the University of Utah's CI, which is closing in June 2023, proffers notable documents or language.

While this report focuses solely on CIs in the United States, it is important to acknowledge that CIs continue to operate in more than 100 countries worldwide (Dig Mandarin, 2021). CIs in some of these nations, including Australia, Austria, Canada, France, Germany, Israel, and the UK, are encountering similar issues as those in the United States as their respective relationships with China deteriorate and as they further contemplate the academic freedom and national

---

[11] See https://www.plu.edu/confucius-institute/agreements/.
[12] Available at https://www.plu.edu/confucius-institute/.
[13] Available at https://www.wesleyancollege.edu/about/confucius-institute.cfm.
[14] Available at https://www.troy.edu/student-life-resources/arts-culture/confucius-institute/about.html.

security implications of hosting foreign-funded entities on university campuses (Today Online, 2022a). In the UK, Prime Minister Rishi Sunak stated prior to his election that he would close all 30 remaining CIs in the UK, since the university-based institutes promote Chinese soft power (Today Online, 2022b; Wilson, 2022). This comes immediately after the September 2022 report released by the Henry Jackson Society, a trans-Atlantic, UK-based think tank, which presents the following series of recommendations regarding CIs in the UK:

- The UK government should introduce legislation to remove CIs from UK universities;
- There must be an immediate investigation regarding the legality of CIs' staff recruitment practices;
- The UK government should add an amendment to the Higher Education (Freedom of Speech) Bill that demands academic partnerships with foreign powers preserve the freedom of speech and comply with equality legislation;
- The UK government should provide £5 million in funding for UK universities to allocate to China studies and bolstering knowledge regarding China's presence in the UK; and
- The UK government should work with other countries that speak Mandarin, especially Taiwan, to develop new Chinese culture and language programmes (Dunning and Kwong, 2022).

ATTRIBUTES

The committee made the following observations during its exploration of CIs at U.S. institutions of higher education:

- CIs are sponsored by the government of the People's Republic of China, which does not protect and promote the core values of U.S. higher education, including the open exchange of ideas and free expression of thought in instruction and scientific research (ChinaFile, 2013).
- The government of the People's Republic of China influences the values, mission, and operational decisions of CIs by being the source of funding; approving the selection of the CI associate/assistant director, and sometimes directly selecting other personnel such as faculty, staff, and instructors; and sometimes maintaining control over operational decisions such as selecting curriculum, texts, programming, and event participants.
- The majority of CIs in the United States are based at colleges and universities (GAO, 2019). CIs are embedded in the life of a campus, as they are physically on, adjacent to, or near their host campuses. CIs welcome students and community members to their curricular and cocurricular

events and activities and often sponsor speakers and visitors, and CI staff participate in campus academic and social life.
- A broad range of U.S. colleges and universities host or have hosted CIs, from major public and private R1 institutions[15] with significant language resources to under-resourced, primarily undergraduate institutions and minority-serving institutions that may depend more on the CI's funding and resources to offer instruction in Chinese language and culture to support their student programming operations.
- CIs at U.S. institutions of higher education are often managed by a board of directors (GAO, 2019). All U.S. institutions appoint their respective CI director, and many individuals appointed to this role have been long-standing members of the faculty and staff of the institution.
- The structure, operations, and management of CIs in the United States varies from institution to institution.
- Each U.S. institution hosting a CI has a corresponding Chinese institution with which it partners and often signs an MOU (CRS, 2022).
- Many CIs hosted at U.S. institutions provide Chinese language instruction and resources to K–12 schools in their region.
  - From an economic development and workforce perspective, state and local governments need young U.S. citizens who are trained in Chinese language and culture and speak Mandarin.
- In some instances, there is a lack of transparency regarding CIs' structure and operations. This can include not providing full, regular disclosure of activities, priorities, and objectives; funding sources and amounts; and bases and processes for decision-making. In addition, many institutions hosting CIs do not post contracts and MOUs publicly and accessibly, and some CIs do not participate in shared university governance processes or structures, which can vary across campuses.

CONTRACTS

Each U.S. institution hosting a CI signs a contract with Hanban to formally establish the institute and to detail funding and other stipulations. Most agreements between U.S. institutions of higher education and Hanban are valid for a 5-year period. In addition, U.S. institutions may establish MOUs or implementation agreements with their respective Chinese partner institutions (GAO, 2019). These MOUs often delineate additional information regarding the structure, management, and activities of the CI.

Many institutions revised their contracts with Hanban, upon renewal or otherwise, after receiving feedback and implementable practices from the U.S.

---

[15] R1 institutions denote universities that offer doctoral degrees and have very high research activity according to the Carnegie Classification System. Additional information is available at https://carnegieclassifications.acenet.edu/classification_descriptions/basic.php.

Department of Education, the American Council on Education, and other higher education associations (Mitchell, 2018). These revisions included affirming the primacy of U.S. law and institutional policies over Chinese law for all CI activities taking place in the United States and removing or revising the original contract's confidentiality language or clauses (GAO, 2019; Mitchell, 2018). For example, Tufts University made changes to its original draft agreement with Hanban following discussions with the National Association of College and University Attorneys to ensure that its CI "complied with Tufts policies, including, without limitation, policies on academic freedom," adding that these revisions included provisions that the CI "be required to comply with Tufts policies and regulations, and that any violation of Tufts policies could result in termination by Tufts of the CI..." (Tufts, 2019).

The Government Accountability Office and the National Association of Scholars have further explored agreements between U.S. institutions of higher education and Hanban (GAO, 2019; Peterson et al., 2022). In addition, the National Association of Scholars maintains a publicly available repository of collected agreements.[16]

RELATIONSHIP OF CONFUCIUS INSTITUTES WITH U.S. CAMPUSES

The structure, operations, and management of CIs in the United States varies from institution to institution (GAO, 2019). In general, a CI has a director or directors, designated CI teachers, and a board of directors. CIs tend to be physically on, adjacent to, or near campus. Some CIs at U.S. institutions are part of an academic department or administrative office, while others are situated elsewhere in the university (GAO, 2019; Kaleem et al., 2022).

The extent to which CIs were integrated into campus life differed by host institution. Some CIs were in prime, widely accessible spaces on campus such as the library, while others were housed in more remote corners (Kaleem et al., 2022). Regardless of where CIs were physically situated, many held cultural events and celebrations open to the campus and the local community.

RELATIONSHIP OF PAIRED CHINESE INSTITUTIONS OF HIGHER EDUCATION WITH U.S. CAMPUSES

The relationship between a U.S. host institution and its Chinese partner institution also varies from institution to institution. Some U.S. institutions intentionally chose their Chinese partner institution, sometimes owing to preexisting relationships, collaborations, or exchanges (Kaleem et al., 2022). For other U.S.

---

[16] See https://drive.google.com/drive/folders/1ozgY69PokmXJMWWO-uBy0DaJ021804iv?usp=sharing.

institutions, Hanban assigned Chinese partner institutions without corresponding U.S. institutions having much, if any, say in the matter (GAO, 2019).

U.S. host institutions approached relationships with their respective Chinese partner institutions in a variety of ways. Contractually, some but not all U.S. host institutions signed supplemental MOUs with their Chinese partner institutions (Kaleem et al., 2022). Some U.S. host institutions saw this partnership, whether formal or informal, as a catalyst, and they leveraged and built upon such relationships with additional language, cultural, and research collaborations. Other U.S. host institutions did not engage extensively with their partner institution beyond matters strictly related to the CI.

3

Benefits and Risks Posed by Confucius Institutes to Academic Institutions

This chapter addresses the benefits of hosting Confucius Institutes (CIs) at U.S. institutions of higher education and the risks that CIs pose to U.S. academic institutions as well as to national security. The task was not to look at all dimensions of CIs, and the following analysis is subject to an important qualification: as the previous chapter notes, the structure, management, and programming of CIs vary widely across institutions. Because of this variation, the committee could not construct a uniform risk profile for all CIs. The chapter has two goals: to provide a comprehensive outline of the benefits and risks a CI might present and to summarize the evidence in the record regarding the experience of institutions that have hosted CIs.

BENEFITS TO U.S. INSTITUTIONS OF HIGHER EDUCATION HOSTING A CONFUCIUS INSTITUTE

Given that U.S. institutions of higher education are under no obligation to host a CI, there must be compelling reasons to establish a CI on a college or university campus. A primary benefit is that a CI provides a source of funds for teaching resources that build current and future capacity in Chinese language and culture at host institutions (Allen-Ebrahimian, 2018; Kaleem et al., 2022). This is particularly true at a time of tight funding for language and area studies at U.S. institutions of higher education and given "a clear need for cultivating Mandarin speakers and China expertise across multiple disciplines" (Horsley, 2021).

For institutions with existing Chinese language programs, these resources can enable them to offer introductory-level courses to more students, provide advanced-level courses that would otherwise be unavailable, and create noncredit

Chinese language courses for members of the local business community (Kaleem et al., 2022). CI instructors can also assist with teacher training and developing teaching materials.[1] In addition, the added resources that come with a CI might enable the host institution to augment cocurricular programming, such as convening conversation groups for students studying Chinese or to meet visiting scholars from China. At some institutions, CI instructors add capacity in teaching courses on Chinese culture and literature as well as language. Collaborations between the host institution and its Chinese partner university have supported research into Chinese language pedagogy.

For institutions that lack their own Chinese language and culture programs, hosting a CI provides opportunities and resources for language instruction on their campuses. CI instructors can also show administrators there is sufficient student interest in learning Chinese and studying Chinese culture to warrant establishing such a program and hiring faculty needed to teach such courses.

Other benefits of hosting a CI include the following:

- Creating opportunities for students to study abroad in China (Tufts University, 2019). This is especially valuable at institutions enrolling many students from lower socioeconomic groups. The COVID-19 pandemic and China's "zero-COVID" policy has limited this opportunity in recent years.
- Providing cultural enrichment experiences on campus and in surrounding communities (Fowler, 2018). CIs have resources to bring in speakers and organize cultural events that provide instructional benefit to students at the U.S. institution and members of the surrounding community.
- Serving as conveners to bring together faculty members whose research relates to China.
- Supporting global engagement and connections. For some universities, it is through the CI that they create a partnership with a Chinese university, which can support other goals such as research collaborations or study abroad (Fowler, 2018). For those institutions whose CI involves an existing partner, the additional activities associated with the CI can strengthen and expand that partnership.
- Offering benefits to the community by providing Chinese language instruction in K–12 schools.
- Serving as a venue for Chinese language proficiency testing services, such as the Hànyǔ Shuǐpíng Kǎoshì (HSK) exam.
- Supporting the internationalization and economic missions of U.S. institutions and their integration into local, state, national, and international businesses. Chinese language instruction and knowledge has become

---

[1] Although host universities do not use Hanban-supplied materials in for-credit classes, these materials can serve as supplementary resources (GAO, 2019).

crucial for deeper understanding of a near-peer competitor and for trade and business between Chinese-speaking countries and the United States.

RISKS TO U.S. INSTITUTIONS OF HIGHER EDUCATION HOSTING A CONFUCIUS INSTITUTE

Academic Freedom and Freedom of Expression

One of the major risks associated with CIs is that they might impair academic freedom on U.S. campuses. This perceived threat relates both to the CI's own activities and to the campus environment more generally.

Academic freedom and freedom of expression and dissent are the foundations of the research, teaching, and learning environment at U.S. colleges and universities. Together, they guarantee that students and faculty members can express their views in free and open intellectual debate and without censorship or sanction. They also guarantee students and faculty the right to pursue academic inquiry on whatever topic they choose and without fear that others will impose their views—including political views—on them.

Interference with the Freedom to Discuss Sensitive Topics

The Chinese government actively censors the flow of information within China on a range of topics, including the status of Taiwan, Uighur rights, and certain historical events such as the Tiananmen Square uprising. One concern regarding CIs is that their presence on U.S. campuses will impair the free and open exchange of information on these topics at the host institution. This concern relates in part to the activities and programs of the CIs themselves. According to the terms of some agreements between CIs and their host institutions, Hanban retains the right to approve proposed expenditures of funds for CI activities and programs. Some observers have expressed doubt that a CI would receive approval for any events involving speakers critical of China's stance on sensitive topics, leading to self-censorship. Others have suggested that CI instructors would avoid such issues within their own classrooms or would stifle any contrary views that emerged during discussion (Diamond and Schell, 2019).

There is also a concern that a CI at an institution of higher education might threaten academic freedom more broadly, with particular impacts on Chinese nationals on campus. Some reports allege that CI staff or board members have attempted either to interfere with university events on politically sensitive topics—by taking down promotional materials, for example—or to pressure universities to avoid such topics altogether (ChinaFile, 2013; Ford, 2022). Other reports identify a more indirect form of interference, suggesting that the mere presence of a CI on campus has a chilling effect on academic discourse by those who fear offending a Chinese partner or the Chinese government (GAO, 2019;

Wallace, 2016). Sixty-eight percent of respondents to the 2018 China Scholar Research Experience Survey "identified self-censorship as a concern for the field" (Chestnut Greitens and Truex, 2018). This concern regarding self-censorship by faculty members or administrators is sometimes ascribed to a fear of losing the funding provided through the CI or other Chinese sources. Finally, students also are affected by self-censorship, as they may choose not to discuss issues sensitive to China or to limit their associations on U.S. campuses out of fear of monitoring by the Chinese Communist Party (FIRE, 2022).

Dissemination of Propaganda

A related concern is that the Chinese government will use CIs to actively disseminate Chinese propaganda, including through the teaching materials that Hanban supplies for use at CIs. This concern is acute at universities with no separate program in Chinese language and culture, which may rely more, or solely, on Hanban-provided materials (Peterson et al., 2022).

Effective Academic Governance

Under the principle of shared governance, it is the joint responsibility of faculty, administrators, and governing bodies to govern U.S. universities (AAUP, 1966). Within this governance model, faculty members play a critical role in overseeing academic matters and participating in decision-making in areas such as faculty appointments and curriculum. However, depending on their particular structure, CIs may sit outside the governance system of the host university, thereby compromising the system of shared governance. The American Association of University Professors has been vocal in raising this concern, stating that "[a]llowing any third-party control of academic matters is inconsistent with principles of ... shared governance" (AAUP, 2014).

Concerns regarding academic oversight of CIs and their operations relate to several matters that the legal agreements among Hanban, the Chinese partner, and the host institution typically address. For example, host universities do not participate in the process by which Hanban screens and selects the teaching staff it recommends for appointment, though U.S. institutions typically have final decision authority as to whether to accept or reject the candidates that Hanban and the Chinese partner university identify. In addition, once recommended, CI instructors might be appointed without going through normal hiring channels subject to faculty involvement and oversight, and once appointed, instructors might not be subject to the control of the relevant academic department or unit (Permanent Subcommittee on Investigations, 2019a). CI administrators might also be exempt from normal employee policies including ongoing performance reviews, and senior administrators might not report up through academic leadership or otherwise be subject to sufficient oversight.

The intersection between these concerns with concerns about a lack of transparency regarding the operations of CIs merits further exploration. For faculty to adequately discharge their shared governance responsibilities, such as executing decision authority over potential CI instructors and determining the nature of course offerings and materials, there should be sufficient transparency in university operations. If confidentiality provisions or other measures make it difficult to obtain information, meaningful shared governance may become impossible. As a result, observers have raised concerns about the privacy surrounding CI legal arrangements (Permanent Subcommittee on Investigations, 2019a).

Finally, the contracts between some U.S. institutions and Hanban contained provisions requiring adherence to both U.S. and Chinese law at CIs on U.S. campuses (Permanent Subcommittee on Investigations, 2019a; Peterson, 2017). This has significant legal and human rights implications and poses an additional concern regarding adequate control over CIs' programs and practices (Gladstone et al., 2021). Many universities hosting CIs subsequently updated their contracts and Memorandums of Understanding to assert the primacy of U.S. law at CIs on U.S. campuses.

University Research

A third set of concerns, in addition to academic freedom and freedom of expression and effective academic governance, relates to university research. This is particularly with respect to the possible effect of CIs on the research environment at host institutions. These concerns overlap with the question of how CIs may affect Department of Defense-funded research, which the following chapter addresses.

Talent Recruitment

China has engaged in an expansive strategy to recruit U.S. researchers. Some observers have noted that CIs provide the Chinese government with a point of access to faculty and students at host institutions (Ren, 2012). This suggests that CI staff may actively engage in recruiting U.S. graduate students and faculty members to contribute to the Chinese research enterprise (Wray, 2018). Possible adverse effects for host universities include losing highly productive researchers and creating conflicts of interest and commitment for faculty members.

Intellectual Property Theft

The majority of CIs do not engage in research,[2] and therefore their programs and activities do not directly intersect with the research environment on U.S. campuses. However, CI staff and faculty interact with members of the university at large and have the potential to serve as operatives of varying capacity for the Chinese government, which actively seeks access to U.S. research data and technology. In addition, some U.S. faculty have obtained permission from Hanban to use CI funding to bring Chinese scholars and artists to their campuses to further their own research as well as joint collaborations. It is possible that some of these visitors to U.S. campuses might also serve as Chinese government operatives.

The committee notes that using CIs as a conduit for intellectual property theft is at best inefficient given the intellectual, and often physical, distance between language and culture studies on campus and departments focusing on scientific research. The public record does not establish that CIs present a special risk in this regard.

Assessment

National Security Risk

Like other forms of activity involving academic interaction with foreign governments of concern, CIs create the potential for threats to research security on campus. However, several committee briefings explicitly addressed this question, and the speakers at these briefings did not view CIs as a primary source of research security threats. Arun Seraphin, deputy director at the National Defense Industrial Association's Emerging Technology Institute, noted on July 20, 2022, that "the 2019 Bipartisan Report led by Senators Portman and Carper found no evidence that CIs are a center for Chinese espionage efforts or any other illegal activity."[3] Kevin Gamache, associate vice chancellor and chief research security officer for the Texas A&M University System, told the committee on July 20, 2022, that CIs "are not a major concern because we no longer have a Confucius Institute within the A&M System. We remain extremely concerned and focus a great deal of effort on understanding and mitigating other state-sponsored programs such as the China Scholarship Council."

As a Brookings Institution report stated, "Multiple investigations into U.S.-based CIs, including by the Senate, have produced no evidence that they facilitate espionage, technology theft or any other illegal activity ..." (Horsley, 2021). This echoes the findings by the Senate Permanent Subcommittee on Investigations

---

[2] Stanford University's CI was a research unit housed within the Department of East Asian Languages and Cultures and conducted academic research on sinology.

[3] The full report is available at www.hsgac.senate.gov/wp-content/uploads/imo/media/doc/PSI%20Report%20China's%20Impact%20on%20the%20US%20Education%20System.pdf.

report mentioned earlier, which states that "... there is no evidence that these [Confucius] institutes are a center for Chinese espionage or any other illegal activity ..." and by Tufts University, which states that while "concerns about Chinese political interference and influence should not be dismissed ... there is not specific evidence regarding CIs" (Permanent Subcommittee on Investigations, 2019b; Tufts, 2019).

Academic Freedom

Based on the public record, it is difficult to determine the extent of the threat CIs present to academic freedom, though Hanban's oversight over programming and budgeting creates this possibility. There are reports of individual instances of problems (Permanent Subcommittee on Investigations, 2019a). For example, in a presentation to the committee on June 22, 2022, Denis Simon, senior adviser to the president for China affairs at Duke University, said, "I would say that everyone involved, unless you were somewhere in outer space during this period, recognized that Chinese money coming from the [People's Republic of China] could not be used to support political programming dealing with Taiwan independence, Tibet independence, human rights, all of the sensitive issues" involving programming at U.S. host institutions. University of Chicago professor Marshall Sahlins, an outspoken critic of CIs, also held this view (Sahlins, 2014). However, the record does not indicate that CIs have had a meaningful impact on free and open discourse on U.S. campuses regarding issues related to China (Abamu, 2019). In fact, several case studies and reports (Diamond and Schell, 2019; GAO, 2019) concluded that the concerns about academic freedom are overstated, and even Simon, in his comments to the committee, acknowledged that "many of the examples of some of this heavy handedness coming from the Chinese side did occur in overseas venues where there were CIs, but I think the conditions that prevailed at U.S. universities, in most cases because they had experience, senior administrators did not allow these things to take place on U.S. campuses."

A 2022 working paper explored the CI teacher selection process and to what extent China's government exerts control over CI teachers (Fan et al., 2022). The authors of this paper found that CI teachers are not screened or selected for their political beliefs or asked to adopt particular political behaviors. CI teachers also receive little training on political topics and encounter minimal monitoring while teaching abroad. However, additional training or political education is unnecessary given the amount of ideological and political education to which Chinese students and scholars are currently exposed. This is consistent with the surveys conducted by Fan, Pan, and Zhang, which revealed that CI teachers frequently espouse and disseminate positions held by the Chinese Communist Party and self-censor when encountering sensitive political topics abroad, as well as Ruth and Xiao, who state that CI teachers "are exceedingly likely to continue to self-censor, regardless of contractual reform, because they know that they will eventually return to China" (Fan et al., 2022; Ruth and Xiao, 2019).

University Governance

The background literature contains more evidence of issues relating to university governance. These include challenges to transparency and, at least prior to universities amending their agreements, gaps in academic oversight and control. One report concluded that universities can and should actively regulate their CIs (Diamond and Schell, 2019).

4

Risks Posed by Confucius Institutes to DOD-Funded Research

The Department of Defense (DOD) research and engineering funding landscape is diverse, in terms of grants and contracts; unclassified, controlled unclassified, and classified programs; and basic, applied, and advanced technology efforts, as well as health, social science, and language programs. This varied and complex landscape requires myriad risk mitigation approaches. While Confucius Institutes (CIs) fundamentally are language and culture institutes, it is imperative to understand the risks that having a CI on, adjacent to, or near campus poses to research, and to DOD-funded research in particular. This chapter seeks to discern whether a CI bestows risks beyond those associated with other foreign individuals, programs, and partnerships already present on campus.

RELATIONSHIP BETWEEN CONFUCIUS INSTITUTES AND DOD-FUNDED RESEARCH

The committee acknowledges that CIs are one aspect of a suite of activities China pursues to engage and influence U.S. academic institutions, which are the bedrock of American education, national security talent, and research. In a June 9, 2022, hearing on U.S.-China competition in global supply chains held by the U.S.-China Economic and Security Review Commission, former U.S. government analyst, linguist, and technology protection expert Jeffrey Stoff testified to the Chinese government's methods of tapping into U.S. innovation engendered by U.S. academic institutions, including "official and unofficial proxies; investment structures such as venture capital funds, incubators and innovation centers; start-up contests; talent programs and supporting recruitment networks; and partnerships with diaspora organizations" (Stoff, 2022).

Maintaining an open environment with a strong commitment to U.S. values, including independence of thought and freedom of speech, is critical. Though the work of this committee is limited by the fact that this study is being conducted at the unclassified level and by its inability to engage with the security community, which was invited but declined to contribute, the committee has not found evidence of espionage or research data theft resulting from the presence of a CI on campus. Therefore, the committee believes that rigorous adherence to appropriate research security measures can mitigate the majority of risk to DOD-funded research posed by a CI on campus and allow DOD to confirm that a CI is not unknowingly integrated into a DOD-funded research program on campus (Rogin, 2018).

RISKS POSED BY CONFUCIUS INSTITUTES TO DOD-FUNDED RESEARCH

The committee believes it is important to be mindful of China's multipronged approach to furthering its national competitiveness and global influence, as China's "hybrid innovation system" blends "forms of academic collaboration, industry partnerships, cyber espionage, direct investment, and influence operations" to further China's power and influence (Fedasiuk et al., 2021; Puglisi, 2020). Specifically, some Chinese "science and technology diplomats" leverage a network of Chinese Communist Party (CCP)–sponsored organizations called the "United Front" to gain intimate details regarding leading-edge science, along with the entities and individuals capable of accelerating China's development (Fedasiuk, 2020; Fedasiuk et al., 2021; Joske, 2020). CIs receive substantial oversight from the United Front Work Department, which reports directly to the CCP Central Committee (Joske, 2020; Xinhua News Agency, 2018).

The CCP can covertly cultivate relationships, build trust, and gain access to strategic information and assets through these nongovernmental but government-organized organizations (Fedasiuk et al., 2021; Lee and Sullivan, 2019). While Chinese Embassy officials responsible for science and technology tend to minimize the role of their overseas influence activities, evidence exists showing that such individuals make payments to overseas scholars and pass funding through Chinese universities (Fedasiuk et al., 2021; Puglisi, 2021). As for the broader role of overseas Chinese professional organizations, more research is needed to illuminate connections between specific groups and China's science and technology ecosystem, particularly regarding military-civil fusion efforts and research parks (Fedasiuk and Weinstein, 2020).

The committee believes it is fair to note that CIs have the potential to be "nontraditional collectors" of foreign technology given their ties to campuses where scientific research is conducted. In fact, Federal Bureau of Investigation Director Christopher Wray and others in the intelligence and security communities perceive CIs in this manner (Redden, 2018; Rogin, 2018; Wray, 2018).

There is evidence that academic solicitation, or "the use of students, professors, scientists or researchers as collectors" of information has increased, tripling from 8 percent of all foreign efforts to obtain classified or sensitive information in fiscal year (FY) 2010 to 24 percent in FY 2014 according to the Defense Security Service (Golden, 2015).

The committee is not aware of any publicly known instances of attempted academic espionage or intellectual property theft stemming from or associated with a CI. However, the committee recognizes that a determined adversary could seek to use any avenue to exploit existing connections and access to research results. While the committee is not aware of any evidence of CIs posing these kinds of risk in the past or at present, the same cannot be said of the future given the evolving political landscape in China. This underscores the need for a rigorous waiver process that contains provisions for increased and sustained oversight of CIs and adequate separation and firewalls between the CI and campus.

The colocation of CIs on campuses hosting classified DOD-funded research is an identified risk. Indeed, CI faculty and staff at a U.S. academic institution might have easier access to campus life and to labs conducting and discussing DOD-funded research. Campuses hosting classified research must adhere to strict requirements issued by the National Industrial Security Program (NISP) and, in particular, those detailed in the National Industrial Security Program Operating Manual (NISPOM).[1] If a campus engages in classified work, the university must have a government sponsor, apply for and receive a facility security clearance, and abide by the rules and regulations required to maintain such clearance. CI-hosting institutions should protect this work or research not only from a CI, but also from any unauthorized access at the university. The committee believes there is no additional risk posed by the conduct of classified research on a campus hosting a CI, as long as the campus complies with NISP and NISPOM requirements.

Another risk could entail having Chinese government representatives tell their own versions of China's culture and traditions, which may conflict with domestic research values including openness and reciprocity. This differs from student organizations that the Chinese government does not authorize or install at a particular university. If the Chinese government directly selects CI faculty, this might be a risk that a CI-hosting institution will need to mitigate because of potential limited transparency regarding the full objectives of the faculty in the U.S. academic institution.

The federal government is currently establishing rules to improve the overall research security posture on campuses through National Security Presidential Memorandum—33 (NSPM-33) (Trump, 2021). These practices appear to cover the risks to DOD-funded research effectively. NSPM-33 addresses and mitigates a broad range of risks by bolstering transparency and reporting requirements. The

---

[1] *National Industrial Security Program Operating Manual*, 32 CFR part 2004. Available at https://www.dcsa.mil/mc/isd/nisp/.

new requirements include full disclosure of outside appointments and activities, including those occurring outside the United States and those conducted outside of the home academic institution's purview. Recent enhancements to other federal agencies' disclosure requirements include listing all foreign appointments, paid and unpaid, as well as all outside resources that support the individual investigator's research program, not only those that support the specific proposed research. As an example, an institution would have to disclose a trainee that a foreign government funds, such as a Chinese scholar funded by the China Scholarship Council, as it would for research materials being shared by a Chinese university. Full knowledge of any investigators' entire research portfolio will permit DOD to make risk-based decisions about whether a particular investigator should participate in a DOD-funded activity.

The committee notes that the majority of institutions that continue to host CIs and to perform DOD-funded research fall below the $50 million threshold of federally funded research that triggers compliance with NSPM-33. Therefore, these campuses must fully comply with all applicable requirements under the federal funding they receive and take it upon themselves to implement strong and effective research security practices in areas including cybersecurity, foreign travel security, insider threat awareness and identification, and export control training. These institutions may find science and security resources developed by the Association of American Universities and Association of Public and Land-grant Universities to be useful (AAU, 2022).

5

Findings

During its discussions and information-gathering activities, the committee identified several risks Confucius Institutes (CIs) pose and developed three categories of findings related to these risks. The first category relates to relevant background and context associated with CIs. The second and third categories focus on the effect of CIs on academic freedom and university governance and on Department of Defense (DOD)–funded research, respectively. The latter two categories allow the committee to expand on overarching concerns regarding CIs: that the presence of a CI on campus may undermine U.S. values of academic freedom, freedom of expression, and academic governance and may jeopardize research security by creating a platform embedded in a U.S. university that the Chinese government could use for nontraditional intelligence gathering and espionage.

FINDINGS REGARDING BACKGROUND AND CONTEXT

1. CIs are one aspect of the Chinese Communist Party's government-coordinated, long-term plan to influence international perspectives about China. Addressing CIs in isolation will not solve the research security challenge on U.S. campuses, and DOD should think strategically about the overall challenge.
2. There is clear evidence of the Chinese government's intent for CIs to promote positive views of China abroad and influence the presentation of issues pertaining to China at institutions of higher education. The CI structure facilitates the Chinese government's direct oversight and influence within CI operations.

3. The U.S. visa-granting system vets all foreign nationals on campus, including CI personnel, graduate students, postdoctoral fellows, and visiting scholars. With this being said, the Chinese government vets all CI staff provided by the Chinese partner institution, and these individuals may pose more of a concern than other Chinese nationals on campus.
4. Many institutions hosting CIs felt that there was real benefit to having the organizations on campus (Fowler, 2018; Kaleem et al., 2022), and that they could maintain the relationship appropriately through increased control, transparency, and clarity in the contractual relationship and management of the CI (Kaleem et al., 2022). This would protect core American university values of openness, transparency, and academic freedom.
5. Federal dollars for language education and area studies have decreased over the past several years, leaving a funding gap and creating an opening for other sources of nonfederal funding, including CIs, to support such activities (Flaherty, 2018; Franks, 2019; Friedman, 2015; NHA, 2022).
6. The loss of CIs on U.S. campuses has had a disproportionate effect on less affluent universities and their communities through the loss of resources and instruction for language and cultural studies at the K–12 and university levels (Allen-Ebrahimian, 2018).
7. Having students who are interested in China and who have received exposure to and training in Chinese language and culture is an economic and national security advantage for the United States in an increasingly complex geopolitical environment (Asia Society, 2005).
8. Industry needs Chinese language and cultural training and programming, as it depends on employees who understand Chinese culture and speak Mandarin to support trade and other international engagements (ACTFL, 2019).
9. DOD eroded trust with universities when it requested that universities submit waiver applications, rejected them all, and then provided no subsequent information about the criteria or reason for refusal (Kaleem et al., 2022). DOD did not issue any waivers in 2018 and 2019 based on its determination that issuing waivers was not in the national interest.

FINDINGS REGARDING THE EFFECT OF CIS ON ACADEMIC FREEDOM AND UNIVERSITY GOVERNANCE

10. CIs create potential risks to academic freedom and freedom of expression on U.S. campuses that disproportionally affect Chinese faculty, staff, and students.

FINDINGS

11. In the past, some CIs have not been subject to the usual academic policies that ensure robust faculty governance and oversight at U.S. institutions of higher education.
12. Some agreements governing the creation and operation of CIs contemplated the application of Chinese law under certain circumstances. This may create conflicts for Chinese faculty, staff, and students.
13. The extent of the risk presented to academic freedom is clear; however, more research is needed to analyze and better understand whether this risk is more significant at smaller institutions with less diversity in sources of learning on Chinese language and culture.

FINDINGS REGARDING THE EFFECT OF CIS ON DOD-FUNDED RESEARCH

14. The committee is not aware of any publicly known instance of attempted espionage associated with a CI, and is not aware of any publicly known evidence that the presence of a CI on campus increases risks to DOD-funded research. However, the committee recognizes that a determined adversary might try to exploit CI connections to access privileged research (Permanent Subcommittee on Investigations, 2019b).
15. CIs, as a foreign government-sponsored entity on campus, pose a risk that host institutions can manage—but not fully eliminate—with policies, procedures, and controls.
16. Good digital and physical security for research is paramount and necessary to protect DOD-funded research, regardless of whether a CI is present on campus (Mroz, 2021).
17. The current processes to protect classified research at institutions of higher education, including compliance with the National Industrial Security Program and the National Industrial Security Program Operating Manual, are sufficient to protect against risks associated with CIs.

6

Recommended Conditions for Granting of a Waiver

In accordance with language found in Section 1062 of the William M. (Mac) Thornberry National Defense Authorization Act (NDAA) for Fiscal Year (FY) 2021,[1] the Statement of Task for this study directs the committee to recommend conditions that a host institution must demonstrate to support a waiver to the limitation on providing funds to institutions of higher education hosting Confucius Institutes (CIs). Congress formulated this language following the waiver process that the Department of Defense (DOD) developed in response to Section 1091 of the John S. McCain National Defense Authorization Act for Fiscal Year 2019.[2]

The DOD waiver process from FY 2019 required universities to provide a substantial amount of information, lacked clear and transparent evaluation criteria, and failed to provide a timeline for considering a waiver application or providing feedback. By not involving key organizations in developing either the waiver criteria or the feedback process in 2018 and 2019, DOD created mistrust with colleges and universities. Ultimately, DOD did not approve any waiver applications following the FY 2019 NDAA, as the agency determined that it was "not in the national interest" to grant waivers (Asimov, 2019). Several institutions felt confident that they would receive a waiver and were genuinely surprised by DOD's decision not to award any (Kaleem et al., 2022). This failure to issue waivers, coupled with a paucity of feedback—which could hamper an institution's ability to reapply and achieve success in the future or to effectively mitigate

---

[1] William M. (Mac) Thornberry National Defense Authorization Act for Fiscal Year 2021, Public Law No. 116-283, 116th Congress, 2nd Session (January 1, 2021), Section 1062 (U.S. Congress, 2021).

[2] John S. McCain National Defense Authorization Act for Fiscal Year 2019, Public Law No. 115-232, 115th Congress, 2nd Session (August 13, 2018) (U.S. Congress, 2018).

real risks that might be present on their campus—led to frustration and a lack of trust in future waiver processes.

A strong and credible waiver process should have as much transparency as possible in balance with legitimate national security concerns and allow institutions of higher education to retain access to resources and opportunities in Chinese language learning and related programs while safeguarding institutions from inappropriate engagements with the Chinese government on U.S. campuses. A potential waiver process, developed with input from key organizations including government, industry, higher education associations, and universities will

- articulate a clear, transparent waiver application process, including a clear data management plan for how this citizen data will be used, who will have access to the data, and how long the data will be maintained;
- include unclassified versions of review and decision protocols along with well-defined criteria for approval;
- identify timelines for review and response; and
- communicate waiver decisions and for denials, provide unclassified information that could be valuable to the institution in addressing national security concerns that could be taking place on their campuses.

The transparency of the waiver process is balanced by the U.S. institutions demonstrating both an understanding of the known risks associated with hosting a CI and that they have taken appropriate steps to mitigate these risks. This is imperative for protecting both DOD-sponsored fundamental and applied scientific research, and university research, academic freedom, and shared governance writ large. The committee notes that the "Agency Waiver Process for [Federal Acquisition Regulation] FAR Prohibition on Covered Telecommunications and Video Surveillance Services or Equipment" that the U.S. Agency for International Development developed in response to Section 889 of the NDAA for FY 2019 may provide a helpful template for organizing and streamlining the recommended waiver criteria into a formal, transparent waiver process (USAID, 2021). The committee also encourages DOD to incorporate aspects of the "Response and Remediation" process employed by the Committee on Foreign Investment in the United States, as this can shift the waiver process from a binary yes or no decision to a constructive, iterative dialogue between DOD and an institution of higher education. The committee wants to be respectful of DOD's bandwidth and believes that such dialogue will not be needed for every waiver application, but rather when circumstances dictate more of a conversation in order to address and mitigate possible concerns.

Finally, the committee recognizes that the U.S. government has the right to withhold taxpayer dollars from going to institutions of higher education where the Chinese government is embedded on campus if proper risk-mitigating measures to secure and protect research data and intellectual property are not present or sufficient.

Recommendation 1.1: Granting Waivers
In the absence of any applicable adverse information that cannot be addressed or mitigated through the criteria below or other means, DOD should grant a waiver if an applying institution of higher education meets the stated waiver criteria.

The committee recognizes that there may be classified reasons why a waiver might not be awarded to an individual institution of higher education. However, in the absence of DOD being aware of adverse information that cannot be addressed or mitigated, and in instances where all other criteria are satisfied, DOD should grant a waiver.

Recommendation 1.2: Communicating about Waivers
If DOD does not grant any waivers, or decides not to grant a waiver to a specific institution of higher education when other waivers are awarded, it should specify the reason(s) for denial to the extent possible at the unclassified level.

This will potentially allow an institution denied a waiver to understand underlying concerns and to address security risks on campus that they may not be aware of. Colleges and universities also may want to consider having access to or retaining a cleared individual in order to have a fuller understanding of security-related issues.

Recommendation 1.3: Establishing the Waiver Application Process
Outside input is critical to ensure that the waiver application process is free from undue administrative and regulatory burden. In addition to U.S. government input, DOD should solicit external input from key organizations, including industry, higher education associations, and universities.

DOD should specify whether this is a one-time, permanent waiver, or provide additional information regarding the duration of the waiver and process for subsequent application, evaluation, and renewal.

Recommendation 2: Waiver Criteria

a. U.S. host institutions should demonstrate that the CI is a formally established Center or Institute at the institution, thereby subjecting the CI to all policies and procedures prescribed in faculty, staff, and student codes, as well as in shared governance documents that ensure that similar units within the university support the key values of American academic institutions, including academic freedom and openness and respectful behavior toward other host institution academic units. If a U.S. host

institution is not structured in a way that allows for formal Centers or Institutes, it should develop a structure for oversight and include the details for that structure in the documents governing a CI.
    i. U.S. host institutions should provide documentation of relevant sections on Centers and Institutes in their institutional handbook or policy manuals.
    ii. U.S. host institutions should provide documentation of an established, regular review process for Centers and Institutes that would ultimately include the CI and would specify the period/frequency of review. This would include external reviews and advisory councils, as required by the institution.
    iii. U.S. host institutions should provide documentation of public statements on academic freedom and freedom of expression as codified in university policy.
    iv. U.S. host institutions should provide documentation of posted CI bylaws and governance documents, including operational and administrative policies and practices.

b. U.S. host institutions should demonstrate that they meet and comply with all applicable DOD requirements for information, data, physical, and research security.
    i. U.S. host institutions conducting $50 million or more of federally funded research per year should demonstrate compliance with National Security Presidential Memorandum – 33 (NSPM-33) or subsequent versions of this document. These institutions can satisfy this criterion by demonstrating that they have established and operate a research security program that includes elements of cybersecurity, foreign travel security, insider threat awareness and identification, and export control training.
    ii. U.S. host institutions conducting less than $50 million of federally funded research per year should demonstrate the presence and implementation of adequate research security measures on campus. These institutions can satisfy this criterion by providing research security-related documents, including elements of cybersecurity, foreign travel security, insider threat awareness and identification, and export control training.
    iii. U.S. host institutions should demonstrate that they have appropriate safeguards in place to ensure that CI faculty and visitors who are not university employees have limited or guest access to university computer networks and cannot access networks that store research results and communications. Institutions can satisfy this criterion by providing the university's cybersecurity and visitor network access

policy and demonstrating that CI faculty and visitors have limited or guest access.
c. U.S. host institutions should demonstrate that they possess full managerial control of CI curriculum, instructors, textbooks and teaching materials, programmatic decisions, and research grants.
   i. U.S. host institutions should demonstrate that CI employees and affiliates are formally associated with the host institution and subject to human resources policies and procedures. Institutions can satisfy this criterion by providing documentation, such as an employment contract or agreement, that the director of the CI is employed by the university with a reporting line to the host institution's chief academic officer or their designee, and by providing public-facing personnel rosters that clearly state whether the host institution classifies CI-affiliated personnel from the Chinese partner institution as either host institution employees or as visiting scholars. CIs should hire their employees and affiliates in accordance with the host institution's human resources policies and procedures and subject to corresponding campus policies.
   ii. U.S. host institutions should demonstrate that CI curricula, including syllabi, textbooks, and teaching materials, are approved through faculty governance review.
   iii. U.S. host institutions should demonstrate oversight over CI-supported research grants. In addition, universities should submit conflict of interest, conflict of commitment, and export control protocols that demonstrate university control.
d. U.S. host institutions should ensure that no contract or other written agreement pertaining to creating or operating the CI calls for the application of foreign law to any aspect of the CI's operation at any U.S. campus of the host institution.
e. U.S. host institutions should demonstrate appropriate fiduciary and financial oversight of the CI. Host institutions can satisfy this criterion by providing the following:
   i. A publicly available annual budget for the CI, including sources of revenue and expenses.
   ii. A copy of the agreement between the host institution and Hanban[3] to host a CI (a hard or electronic copy of the original document, including the original version of the agreement in Chinese). The host institution should manage the agreement through its sponsored program process.

---

[3] "Hanban" is the colloquial term for the Chinese International Education Foundation, or CIEF, now known as the Ministry of Education Center for Language Education and Cooperation, or CLEC. This is the Chinese government agency affiliated with China's Ministry of Education that promoted, managed, and funded CIs on foreign campuses.

iii. A copy of the Memorandum of Understanding and contract between the U.S. host institution and Chinese partner institution, if applicable.
iv. A copy of the policy that any financial contribution from foreign or domestic sources supporting the CI must be treated as a sponsored contract, not a gift, with a deliverable (programming, education, etc.).

The committee hopes that the recommendations and waiver criteria it proposes in this report are useful to DOD as the agency works to craft a fair, credible, and implementable waiver process in the coming months. The committee looks forward to undertaking additional research to understand foreign-funded programs and entities on campus more broadly and to identify implementable practices and principles regarding appropriate operations for academic institutions involving foreign partnerships in the coming months. The committee will present these, along with additional findings and recommendations, in a second report to be released in June 2023.

References

AAU (Association of American Universities). 2022. *Science and Security Resources*. Retrieved from https://www.aau.edu/sites/default/files/AAU-Files/Key-Issues/Science-Security/Science-and-Security-Resource-Document.pdf.

AAUP (American Association of University Professors). 1966. *Statement on Government of Colleges and Universities*. Retrieved from https://www.aaup.org/report/statement-government-colleges-and-universities.

AAUP. 2014. *On Partnerships with Foreign Governments: The Case of Confucius Institutes*. Retrieved from https://www.aaup.org/file/Confucius_Institutes_0.pdf.

Abamu, J. 2019. University of Maryland faces questions over Chinese government's role in program. *WAMU 88.5*, August 22, 2019. Retrieved from https://wamu.org/story/19/08/22/university-of-maryland-faces-questions-over-chinese-governments-role-in-program/.

ACTFL (American Council on the Teaching of Foreign Languages). 2019. *Making Languages Our Business: Addressing Foreign Language Demand Among U.S. Employers*. Retrieved from https://www.leadwithlanguages.org/wp-content/uploads/MakingLanguagesOurBusiness_Full-Report.pdf.

Albert, E., L. Maizland, and B. Xu. 2021. *The Chinese Communist Party*. Washington, DC: Council on Foreign Relations. Retrieved from https://www.cfr.org/backgrounder/chinese-communist-party.

Allen-Ebrahimian, B. 2018. How China managed to play censor at a conference on U.S. soil. *Foreign Policy*, May 9, 2018. Retrieved from https://foreignpolicy.com/2018/05/09/how-china-managed-to-play-censor-at-a-conference-on-u-s-soil/.

Asia Society. 2005. *Expanding Chinese-Language Capacity in the United States: What Would It Take to Have 5 Percent of High School Students Learning Chinese by 2015?* Retrieved from https://asiasociety.org/files/expandingchinese_0.pdf.

Asimov, N. 2019. SFSU shutters popular Chinese cultural program under pressure from feds. *San Francisco Chronicle*, August 5, 2019. Retrieved from https://www.sfchronicle.com/bayarea/article/SFSU-shutters-popular-Chinese-cultural-program-14278818.php.

Brady, A.-M. 2017. *Magic Weapons: China's Political Influence Activities Under Xi Jinping*. Washington, DC: The Wilson Center, *Insight & Analysis*. Retrieved from https://www.wilsoncenter.org/article/magic-weapons-chinas-political-influence-activities-under-xi-jinping.

Bryant University. 2021. Bryant declines to apply for continued funding for Confucius Institute. Press release, April 6, 2021. Retrieved from https://news.bryant.edu/bryant-declines-apply-continued-funding-confucius-institute.

Bullard, G. 2020. Following federal pressure, UMD will close a program that had Chinese government support. *WAMU 88.5*, January 17, 2020. Retrieved from https://wamu.org/story/20/01/17/following-federal-pressure-umd-will-close-a-program-that-had-chinese-government-support/.

Chestnut Greitens, S., and R. Truex. 2018. Repressive experiences among China scholars: New evidence from survey data. *SSRN* (August 1). Retrieved from https://ssrn.com/abstract=3243059.

ChinaFile. 2013. *How Much Is a Hardline Party Directive Shaping China's Current Political Climate?* Document 9: A *ChinaFile* Translation, November 8, 2013. Retrieved from https://www.chinafile.com/document-9-chinafile-translation.

CIWA (Confucius Institute of the State of Washington). 2022. Pacific Lutheran University (department website). Retrieved from https://www.plu.edu/confucius-institute/.

CRS (Congressional Research Service). 2022. *Confucius Institutes in the United States: Selected Issues*. Retrieved from https://crsreports.congress.gov/product/pdf/IF/IF11180.

Diamond, L., and O. Schell. 2019. *China's Influence and American Interests: Promoting Constructive Vigilance*. Stanford, CA: Hoover Institution Press. Retrieved from https://www.hoover.org/research/chinas-influence-american-interests-promoting-constructive-vigilance.

Dig Mandarin. 2021. Confucius Institutes Around the World – 2021. Retrieved from https://www.digmandarin.com/confucius-institutes-around-the-world.html.

DOJ (U.S. Department of Justice). 2021. Information About the Department of Justice's China Initiative and a Compilation of China-related Prosecutions Since 2018. Retrieved from https://www.justice.gov/archives/nsd/information-about-department-justice-s-china-initiative-and-compilation-china-related.

Dunning, S., and A. Kwong. 2022. *An Investigation of China's Confucius Institutes in the UK*. London: Henry Jackson Society. Retrieved from https://henryjacksonsociety.org/publications/an-investigation-of-chinas-confucius-institutes-in-the-uk/.

Fan, Y., J. Pan, and T. Zhang. 2022. Confucius Institutes: Vehicles of CCP Propaganda? *SCCEI China Briefs*. Stanford, CA: Stanford University, Center on China's Economy and Institutions. Retrieved from https://sccei.fsi.stanford.edu/content/confucius-institutes-vehicles-ccp-propaganda.

Fedasiuk, R. 2020. Putting money in the party's mouth: How China mobilizes funding for United Front work. *China Brief* 20(16). Retrieved from https://jamestown.org/program/putting-money-in-the-partys-mouth-how-china-mobilizes-funding-for-united-front-work/.

Fedasiuk, R., and E. Weinstein. 2020. *Overseas Professional and Technology Transfer to China*. Washington, DC: Center for Security and Emerging Technology. Retrieved from https://cset.georgetown.edu/wp-content/uploads/CSET-Overseas-Professionals-and-Technology-Transfer-to-China.pdf.

Fedasiuk, R., E. Weinstein, and A. Puglisi. 2021. *China's Foreign Technology Wish List*. Washington, DC: Center for Security and Emerging Technology. Retrieved from https://cset.georgetown.edu/wp-content/uploads/CSET-Chinas-Foreign-Technology-Wish-List.pdf.

Feldgoise, J., and R. Zwetsloot. 2022. *Estimating the Number of Chinese STEM Students in the United States*. Washington, DC: Center for Security and Emerging Technology. Retrieved from https://cset.georgetown.edu/publication/estimating-the-number-of-chinese-stem-students-in-the-united-states/.

Financial Times. 2022. America is struggling to counter China's intellectual property theft. Retrieved from https://www.ft.com/content/1d13ab71-bffd-4d63-a0bf-9e9bdfc33c39.

FIRE (Foundation for Individual Rights and Expression). 2022. *Tracker: University Responses to Chinese Censorship*. Retrieved from https://www.thefire.org/resources/home-abroad-resources/universities-respond-to-chinas-censorship-efforts/.

Flaherty, C. 2018. L'œuf ou la poule? *Inside Higher Ed* (March 19). Retrieved from https://www.insidehighered.com/news/2018/03/19/mla-data-enrollments-show-foreign-language-study-decline.

REFERENCES

Flaherty, C. 2019. Blowback for Huawei bans. *Inside Higher Ed* (May 31). Retrieved from https://www.insidehighered.com/news/2019/05/31/researchers-want-less-restrictive-policies-accepting-money-chinese-telecoms-giant.

Ford, W. 2022. How far does China's influence at U.S. universities go? One student tried to find out. *Politico* (April 24). Retrieved from https://www.politico.com/news/magazine/2022/04/24/confucius-institutes-china-new-mexico-00027287.

Fowler, V. L. 2018. Dear leaders and friends of higher education. Press release, March 21, 2018. Wesleyan College. Retrieved from https://www.wesleyancollege.edu/academics/upload/openletter.pdf.

Franks, S. 2019. America's failure to fund language education is creating a crisis. Language-Line Solutions blog, posted March 4, 2019. Retrieved from https://blog.languageline.com/americas-failure-to-fund-language-education-is-creating-a-crisis.

Friedman, A. 2015. America's lacking language skills. *The Atlantic* (May 10). Retrieved from https://www.theatlantic.com/education/archive/2015/05/filling-americas-language-education-potholes/392876/.

Friis, K., and O. Lysne. 2021. Huawei, 5G and security: Technological limitations and political responses. *Development and Change* 52(5), 1174–1195. DOI: https://doi.org/10.1111/dech.12680.

GAO (Government Accountability Office). 2019. *Agreements Establishing Confucius Institutes at U.S. Universities Are Similar, But Institute Operations Vary* (GAO-19-278). Retrieved from https://www.gao.gov/assets/gao-19-278.pdf.

Gibney, E. 2019. UC Berkeley bans new research funding from Huawei. *Nature* 566(7742), 16–17. DOI:10.1038/d41586-019-00451-z. Retrieved from https://www.nature.com/articles/d41586-019-00451-z.

Gladstone, Z., J. Ho, and J. Wang. 2021. *Unraveling China's Attempts to Hinder Academic Freedom: Confucius Institutes*. New York: Human Rights Foundation. Retrieved from https://hrf.org/unraveling-chinas-attempts-to-hinder-academic-freedom-confucius-institutes/.

Golden, D. 2015. Testimony to the House Science Committee by Daniel Golden, April 11, 2015 (U.S. House of Representatives, Committee on Science, Space, and Technology). Retrieved from https://docs.house.gov/meetings/SY/SY21/20180411/108175/HHRG-115-SY21-Wstate-GoldenD-20180411.pdf.

Green-Riley, N. 2020. The State Department labeled China's Confucius programs a bad influence on U.S. students. What's the story? *The Washington Post*, August 25, 2020. Retrieved from https://www.washingtonpost.com/politics/2020/08/24/state-department-labeled-chinas-confucius-programs-bad-influence-us-students-whats-story/.

Horsley, J. P. 2021. It's time for a new policy on Confucius institutes. *Brookings Institution*, April 1, 2021. Retrieved from https://www.brookings.edu/articles/its-time-for-a-new-policy-on-confucius-institutes/.

JASON. 2019. *Fundamental Research Security*. McLean, VA: JASON, The MITRE Corporation. Retrieved from https://www.nsf.gov/news/special_reports/jasonsecurity/JSR-19-2IFundamentalResearchSecurity_12062019FINAL.pdf.

Joske, A. 2020. *The Party Speaks for You: Foreign Interference and the Chinese Communist Party's United Front System*. Barton, Australia: Australian Strategic Policy Institute. Retrieved from https://www.aspi.org.au/report/party-speaks-you.

Kaleem, K., J. Howard, and K. Macdonald. 2022. *Information-Gathering Sessions with U.S. Colleges and Universities That Are Current or Former Hosts of Confucius Institutes*. Arlington, VA: American Institutes for Research.

Knowles, H., and B. Jin. 2019. Warnings of Chinese government 'influence' on campuses divide Stanford community. *The Stanford Daily*, May 30, 2019. Retrieved from https://stanforddaily.com/2019/05/30/warnings-of-chinese-government-influence-on-campuses-divide-stanford-community/.

Lee, K., and A. Sullivan. 2019. *People's Republic of the United Nations: China's Emerging Revisionism in International Organizations*. Washington, DC: Center for a New American Security. Retrieved from https://www.cnas.org/publications/reports/peoples-republic-of-the-united-nations.

Ma, Y. 2020. *Ambitious and Anxious: How Chinese College Students Succeed and Struggle in American Higher Education*. New York: Columbia University Press.

Mitchell, T. 2018. Letter to the President of the United States. Washington, DC: American Council on Education. Retrieved from https://www.acenet.edu/Documents/Letter-on-Confucius-Institutes.pdf.

Mosher, S. W. 2012. Confucius Institutes: Trojan Horses with Chinese Characteristics. Testimony Presented to the Subcommittee on Oversight and Investigations House Committee on Foreign Affairs, March 28, 2012. Retrieved from https://www.iwp.edu/wp-content/uploads/2019/05/20181017_MosherConfuciusInstitutes.pdf.

Mroz, I. 2021. *Understanding Higher Education Cybersecurity Threats to Research and IP*. archTIS blog, posted December 2, 2021. Retrieved from https://www.archtis.com/understanding-higher-education-cybersecurity-threats-to-research-and-ip/.

NBC 10 WJAR News. 2021. Bryant University will not seek funding for the Confucius Institute. March 22, 2021. Retrieved from https://turnto10.com/news/local/bryant-university-will-not-seek-funding-for-the-confucius-institute-after-15-years.

NCES (National Center for Education Statistics). 2022. Undergraduate Enrollment. *Condition of Education* (web page). Washington, DC: U.S. Department of Education, Institute of Education Sciences. Retrieved from https://nces.ed.gov/programs/coe/indicator/cha.

NHA (National Humanities Alliance). 2022. Huge cuts proposed to already depleted international education and foreign language programs. National Humanities Alliance blog, n.d. Retrieved from https://www.nhalliance.org/titlevi_cuts.

NSB (National Science Board, National Science Foundation). 2022. *Science and Engineering Indicators 2022: The State of U.S. Science and Engineering*. NSB-2022-1. Retrieved from https://ncses.nsf.gov/pubs/nsb20221.

NSCRC (National Student Clearing House Research Center). 2022. *Overview: Spring 2022 Current Term Enrollment Estimates*. Retrieved from https://nscresearchcenter.org/current-term-enrollment-estimates/.

NSF (National Science Foundation). 2021. *Higher Education Research and Development: Fiscal Year 2020*. NSF 22-311. Retrieved from https://ncses.nsf.gov/pubs/nsf22311.

Obama, B. 2010. Executive Order 13556—Controlled Unclassified Information. The White House, Office of the Press Secretary, November 4, 2010. Washington, DC: Executive Office of the President of the United States. Retrieved from https://obamawhitehouse.archives.gov/the-press-office/2010/11/04/executive-order-13556-controlled-unclassified-information.

OpenDoors. 2021. *2021 Fact Sheet: China*. Retrieved from https://opendoorsdata.org/wp-content/uploads/2021/11/Country-Sheet_China_2021.pdf.

Padilla, J. D. 2021. Closing the Confucius Institute at Valparaiso University. Press release, August 30, 2021. Retrieved from https://www.valpo.edu/news/2021/08/30/closing-the-confucius-institute-at-valparaiso-university-civu/.

Permanent Subcommittee on Investigations. 2019a. China's Impact on the U.S. Education System. Staff Report. Washington, DC: U.S. Senate. Retrieved from https://www.hsgac.senate.gov/wp-content/uploads/imo/media/doc/PSI%20Report%20China's%20Impact%20on%20the%20US%20Education%20System.pdf.

Permanent Subcommittee on Investigations. 2019b. Senators Portman & Carper unveil bipartisan report on Confucius Institutes at U.S. universities & K-12 classrooms. Media release, February 27, 2019. Retrieved from https://www.hsgac.senate.gov/subcommittees/investigations/media/senators-portman-and-carper-unveil-bipartisan-report-on-confucius-institutes-at-us-universities_k-12-classrooms.

REFERENCES

Peterson, R. 2017. *Outsourced to China: Confucius Institutes and Soft Power in American Higher Education*. New York: National Association of Scholars. Retrieved from https://www.nas.org/storage/app/media/Reports/Outsourced to China/NAS_confuciusInstitutes.pdf.

Peterson, R., F. Yan., and I. Oxnevad. 2022. *After Confucius Institutes: China's Enduring Influence on American Higher Education*. New York: National Association of Scholars. Retrieved from https://www.nas.org/storage/app/media/Reports/After%20Confucius%20Institutes/After_Confucius_Institutes_NAS.pdf.

Puglisi, A. 2020. The Myth of the Stateless Global Society. In W. C. Hannas and D. K. Tatlow (eds.), *China's Quest for Foreign Technology*. London: Routledge.

Puglisi, A. 2021. Beijing's Long Arm: Threats to U.S. National Security. Testimony before the Senate Select Committee on Intelligence, August 4, 2021. Washington, DC: Center for Security and Emerging Technology. Retrieved from https://cset.georgetown.edu/publication/anna-puglisis-testimony-before-the-senate-select-committee-on-intelligence/.

Redden, E. 2012. Confucius says... *Inside Higher Ed* (January 4). Retrieved from https://www.insidehighered.com/news/2012/01/04/debate-over-chinese-funded-institutes-american-universities#ixzz1qBcE0hjA.

Redden, E. 2017. New Scrutiny for Confucius Institutes. *Inside Higher Ed* (April 26). Retrieved from https://www.insidehighered.com/news/2017/04/26/report-confucius-institutes-finds-no-smoking-guns-enough-concerns-recommend-closure.

Redden, E. 2018. The Chinese Student Threat? *Inside Higher Ed* (February 15). Retrieved from https://www.insidehighered.com/news/2018/02/15/fbi-director-testifies-chinese-students-and-intelligence-threats.

Ren, Z. 2012. The Confucius Institutes and China's Soft Power. IDE Discussion Papers 330. Institute of Developing Economies, Japan's External Trade Organization (JETRO). Retrieved from https://ideas.repec.org/p/jet/dpaper/dpaper330.html.

Riley, C. 2013. UI avoids controversy with Confucius Institute contract. *The Daily Iowan*, February 28, 2013. Retrieved from https://dailyiowan.com/2013/02/28/ui-avoids-controversy-with-confucius-institute-contract/.

Rogin, J. 2018. Preventing Chinese espionage at American universities. *The Washington Post* (May 22, 2018). Retrieved from https://www.washingtonpost.com/news/josh-rogin/wp/2018/05/22/preventing-chinese-espionage-at-americas-universities/.

Ross, R., V. Pillitteri, K. Dempsey, M. Riddle, and G. Guissanie. 2020. *Protecting Controlled Unclassified Information in Nonfederal Systems and Organizations*. Gaithersburg, MD: National Institute of Standards and Technology. Retrieved from https://doi.org/10.6028/NIST.SP.800-171r2.

Rui, G. 2022. China's Communist Party nears 97 million, with more younger and educated members. *South China Morning Post*, November 30, 2022. Retrieved from https://www.scmp.com/news/china/politics/article/3183669/chinas-communist-party-grows-near-97-million-its-made-younger.

Ruth, J., and Y. Xiao. 2019. *Academic Freedom and China*. Washington, DC: American Association of University Professors. Retrieved from https://www.aaup.org/article/academic-freedom-and-china#.Y2lzWXbMI2w.

Sabrowsky, H. 2019. China-funded institute set to close. *The Minnesota Daily*, February 21, 2019. Retrieved from https://mndaily.com/242511/news/adconfucius/.

Sahlins, M. 2014. *Confucius Institutes*. Chicago: Prickly Paradigm Press, LLC.

Schmidt, C. 2014. University to end partnership with Confucius Institute. *The Chicago Maroon*, September 30, 2014. Retrieved from http://chicagomaroon.com/19307/news/university-to-end-partnership-with-confucius-institute/.

Stoff, J. 2022. Reassessing Threats to US Innovation Posed by China and Implications for Safeguarding Future Supply Chains. Testimony Before the U.S.- China Economic and Security Review Commission. Hearing on "U.S.–China Competition in Global Supply Chains," June 9, 2022. Retrieved from https://www.uscc.gov/sites/default/files/2022-06/Jeff_Stoff_Testimony.pdf.

Subcommittee on Research Security, Joint Committee on the Research Environment. 2022. *Guidance for Implementing National Security Presidential Memorandum 33 (NSPM-33) on National Security Strategy for United States Government-Supported Research and Development.* Washington, DC: National Science and Technology Council. Retrieved from https://www.whitehouse.gov/wp-content/uploads/2022/01/010422-NSPM-33-Implementation-Guidance.pdf.

Thennarasu, E. 2019. URI cuts ties with Confucius Institute. *The Good Five Cent Cigar*, February 7, 2019. Retrieved from https://rhodycigar.com/2019/02/07/dissolution-of-the-confucius-institute/.

Today Online. 2022a. 'Malign influence': China's cultural institutes under growing scrutiny. October 9, 2022. Retrieved from https://www.todayonline.com/world/malign-influence-chinas-cultural-institutes-under-growing-scrutiny-2013886.

Today Online. 2022b. UK PM hopeful Sunak takes aim at China in leadership contest. July 25, 2022. Retrieved from https://www.todayonline.com/world/uk-pm-hopeful-sunak-takes-aim-china-leadership-contest-1952541.

Trump, D. 2021. Presidential Memorandum on United States Government-Supported Research and Development National Security Policy. National Security Presidential Memorandum – 33, January 14, 2021. Washington, DC: The White House. Retrieved from https://trumpwhitehouse.archives.gov/presidential-actions/presidential-memorandum-united-states-government-supported-research-development-national-security-policy/.

Tufts (Tufts University). 2019. *Findings of the Confucius Institute Review Committee and Decision of the University Administration Regarding the Confucius Institute at Tufts University.* Retrieved from https://provost.tufts.edu/wp-content/uploads/CITU-report-for-publication-final-clean.pdf.

Tufts (Tufts University). 2021. Decision to close the Confucius Institute at Tufts University. Office of the Provost and Senior Vice President blog, posted March 17, 2021. Retrieved from https://provost.tufts.edu/blog/news/2021/03/17/decision-to-close-the-confucius-institute-at-tufts-university/.

University of Akron News. 2021. UA to close Confucius Institute. Press release, November 8, 2021. Retrieved from https://www.uakron.edu/im/news/ua-to-close-confucius-institute.

USAID (U.S. Agency for International Development). 2021. *Agency Waiver Process for FAR Prohibition on Covered Telecommunications and Video Surveillance Services or Equipment: A Mandatory Reference for ADS Chapter 302.* Retrieved from https://www.usaid.gov/ads/policy/300/302mbp.

U.S. Congress. 2018. H.R. 5515 – John S. McCain National Defense Authorization Act for Fiscal Year 2019. Retrieved from https://www.congress.gov/bill/115th-congress/house-bill/5515.

U.S. Congress. 2021. H.R. 6395 – William M. (Mac) Thornberry National Defense Authorization Act for Fiscal Year 2021. Retrieved from https://www.congress.gov/bill/116th-congress/house-bill/6395.

U.S. Congress. 2022. H.R.4346 – Supreme Court Security Funding Act of 2022. Retrieved from https://www.congress.gov/bill/117th-congress/house-bill/4346.

Wallace, R. L. 2016. *The Confucius Institutes in the Real World.* B.Ph. thesis, University of Pittsburgh. Retrieved from http://d-scholarship.pitt.edu/27901/1/The_Confucius_Institutes_in_the_Real_World.pdf.

Weinstein, E. 2022. *Chinese Talent Program Tracker.* Washington, DC: Center for Security and Emerging Technology. Retrieved from https://cset.georgetown.edu/publication/chinese-talent-program-tracker/.

Wilson, N. 2022. Rishi Sunak 'looking to close' Confucius Institutes across UK universities. *The National*, November 1, 2022. Retrieved from https://www.thenational.scot/news/23094268.rishi-sunak-looking-close-confucius-institutes-pose-threat-civil-liberties/.

World Bank. 2022. Population, total – China. Retrieved from https://data.worldbank.org/indicator/SP.POP.TOTL?locations=CN.

Wray, C. 2018. Testimony to the U.S. Senate Select Committee on Intelligence, February 13, 2018. Retrieved from https://www.intelligence.senate.gov/hearings/open-hearing-worldwide-threats-0#.

REFERENCES

Xinhua News Agency. 2018. Sun Chunlan emphasized at the 13th Confucius Institute Conference to promote the high-quality development of Confucius Institutes and contribute to building a community with a shared future for mankind. Press release, December 4, 2018. Retrieved from https://web.archive.org/web/20190902020727/http:/www.hanban.org/article/2018-12/04/content_754703.htm.

Zialcita, P. 2021. CSU will close its Confucius Institute rather than risk loss of federal funding. *Colorado Public Radio News*, March 6, 2021. Retrieved from https://www.cpr.org/2021/03/06/csu-will-close-its-confucius-institute-rather-than-risk-loss-of-federal-funding/.

Appendix A

Committee Biographical Information

PHILIP J. HANLON (CHAIR)

Philip J. Hanlon is president and professor of mathematics at Dartmouth College. He is an experienced academic leader having served as provost at the University of Michigan prior to assuming his current role. Dr. Hanlon has been an active leader within the U.S. academy and is currently serving as chair of the COFHE Board of Directors and chair of the Ivy President's Council. For the past 5 years, Dr. Hanlon has served the National Academies of Sciences, Engineering, and Medicine as a member of the Policy and Global Affairs Committee. As a mathematician, his research interests are in algebraic combinatorics and discrete probability. During his academic career he has received numerous awards for his research and teaching, including a Presidential Young Investigator Award, a Guggenheim Fellowship, a Sloan Fellowship, a Thurnau Professorship at the University of Michigan, as well as membership in the American Academy of Arts and Sciences. He received a B.A. from Dartmouth College and a Ph.D. in mathematics from the California Institute of Technology.

JAYATHI Y. MURTHY (VICE CHAIR)

Jayathi Y. Murthy is president and professor of mechanical, industrial, and manufacturing engineering at Oregon State University. She previously served as the Ronald and Valerie Sugar Dean of the Henry Samueli School of Engineering and Applied Science at the University of California, Los Angeles. Dr. Murthy has worked in both industry and academia, including a decade as one of the early employees of Fluent Inc., a leading vendor of computational fluid dynamics software. She directed the National Nuclear Security Administration's Center for

Prediction of Reliability, Integrity, and Survivability of Microsystems, or PRISM, between 2008 and 2014. Dr. Murthy's expertise is in computational fluid dynamics and heat transfer, and most recently, her work has focused on submicron thermal transport and uncertainty quantification in multiscale multiphysics systems. She is the author of more than 300 technical publications and is the recipient of numerous awards and recognitions, including the ASME Heat Transfer Memorial Award in 2016. She was inducted into the National Academy of Engineering in 2020. Dr. Murthy received her Ph.D. from the University of Minnesota, her M.S. from Washington State University, and a B.Tech. from the Indian Institute of Technology, Kanpur, India.

HANNAH L. BUXBAUM

Hannah L. Buxbaum is vice president for international affairs at Indiana University (IU). She is also professor of law and John E. Schiller Chair at IU's Maurer School of Law, where she previously held leadership roles, including as interim dean (2012–2014). As vice president, she advances global engagement across IU's seven campuses, including at the Bloomington campus, which hosts 3 Department of Defense–funded Language Flagship Programs (Arabic, Chinese, Russian) and 18 Department of Education Title VI programs, 8 of which are designated as National Resource Centers. She oversees the offices that manage international admissions and student services, study abroad, international partnerships, and international development, as well as the university's five Global Gateway offices. Ms. Buxbaum serves on the Fulbright Scholar CIES Advisory Board and the Association of International Education Administrators' Public Policy Committee, and is currently chair of the Big Ten Academic Alliance's Senior International Officer group. Ms. Buxbaum is an elected member of the American Law Institute and the International Academy of Comparative Law, and in 2019 she was appointed as the U.S. member of the Curatorium of The Hague Academy of International Law. She holds a B.A. from Cornell University, a J.D. from Cornell Law School, and an LL.M. from the University of Heidelberg.

CLAUDE R. CANIZARES

Claude R. Canizares is the Bruno Rossi Professor of Physics at the Massachusetts Institute of Technology (MIT). He has served as vice president (2013–2015), vice president for research and associate provost (2006–2013), associate provost (2001–2006), and director of the Center for Space Research (1990–2002). He oversaw MIT Lincoln Laboratory and MIT's international engagements. He is a principal investigator on NASA's Chandra X-ray Observatory and studies X-ray sources, including active stars, black-hole or neutron star binaries, supernova remnants, quasars, and clusters of galaxies. His service outside MIT has included the Department of Commerce's National Advisory Council on Innovation and Entrepreneurship and the National Academies of Sciences, Engineering, and Medicine's Committees on Science, Technology, and Law; Science, Security, and Prosperity; and Federal Regulations and Reporting Requirements. He served as chair of the National Academies' Space Studies Board and was a member of the NASA Advisory Council, the Air Force Scientific Advisory Board, and the board of Associated Universities, Inc., among others. He served as a member of the L-3 Technologies, Inc. (now L3Harris) Board of Directors (2003–2019). He currently serves on the Hubble Space Telescope Institute Council and the National Academies' Air Force Studies Board, and chairs the Audit Committee of the National Academy of Sciences, among others. Dr. Canizares is a member of the National Academy of Sciences and the International Academy of Astronautics and is a fellow of the American Academy of Arts and Sciences, the American Physical Society, and the American Association for the Advancement of Science. His awards include Decoration for Meritorious Civilian Service to the United States Air Force, the Goddard medal, the Basic Sciences Award of the International Academy of Astronautics, and two NASA Public Service Medals.

ROBERT L. DALY

Robert L. Daly directs the Kissinger Institute on China and the United States at the Wilson Center. He previously directed the University of Maryland China Initiative and served as American Director of the Johns Hopkins University–Nanjing University Center for Chinese and American Studies in Nanjing. He began work in U.S.-China relations as Cultural Exchanges Officer at the U.S. Embassy in Beijing. After leaving the Foreign Service, he taught Chinese at Cornell University, worked on television and theater projects in China, and helped produce the Chinese-language version of Sesame Street. He is a director of the National Committee on U.S.-China Relations and American Mandarin Society and a member of the Task Force on U.S. China Policy. Mr. Daly has testified before Congress, and his analysis is featured on NPR, C-SPAN, CNBC, and the Voice of America. He has interpreted for Chinese and American leaders, including Jiang Zemin and Henry Kissinger, and has lived in China for 12 years.

PETER K. DORHOUT

Peter K. Dorhout serves as professor of chemistry and vice president for research at Iowa State University (ISU) and is an Ames Laboratory affiliate. Prior to ISU, he served 5 years as vice president for research following 4 years as dean of the College of Arts and Sciences at Kansas State University. He served 20 years at Colorado State University as vice provost for graduate studies, interim director of international programs, and professor of chemistry. He has served as a collaborator at Los Alamos National Laboratory since 1987. He has led professional organizations as a member of the Boards of Directors for the American Chemical Society, where he was the 2018 president, and the Research Corporation for Science Advancement. He is a recognized expert in solid state and nuclear materials chemistry as well as a project advisor with the Russian Federal Nuclear Center through the CRDF Global. He earned a B.S. in chemistry from the University of Illinois at Urbana-Champaign and a Ph.D. in chemistry from the University of Wisconsin–Madison, and served as a postdoctoral scientist at Ames Laboratory at ISU. Dr. Dorhout is a fellow of the American Chemical Society and the American Association for the Advancement of Science and is a Research Corporation for Science Advancement Cottrell Scholar. His awards include a Sloan Fellowship and the ACS-ExxonMobil Faculty Award in Solid State Chemistry.

MELISSA L. FLAGG

Melissa L. Flagg is the founder of Flagg Consulting LLC, as well as a fellow at the Acquisition Innovation Research Center (AIRC), a visiting fellow at the Perry World House at the University of Pennsylvania, and a senior advisor to the Atlantic Council GeoTech Center. Prior to this, she was a senior fellow at the Center for Security and Emerging Technology at Georgetown University. Previously she served as the deputy assistant secretary of defense for research, responsible for policy and oversight of Defense Department science and technology programs. She has worked at the State Department, the Office of Naval Research, the Office of the Secretary of Defense for Research and Engineering, the John D. and Catherine T. MacArthur Foundation, and the Army Research Laboratory. Dr. Flagg has served on numerous boards, including the National Academy of Sciences Air Force Studies Board (2014–2015) and the Department of Commerce Emerging Technology Research Advisory Committee. She holds a Ph.D. in pharmaceutical chemistry and a B.S. in pharmacy.

MARY GALLAGHER

Mary E. Gallagher is the Amy and Alan Lowenstein Professor of Democracy, Democratization, and Human Rights at the University of Michigan, where she is also the director of the International Institute. She was the director of the Kenneth G. Lieberthal and Richard H. Rogel Center for Chinese Studies from 2008 to 2020. Dr. Gallagher's most recent book is *Authoritarian Legality in China: Law, Workers, and the State*, published by Cambridge University Press in 2017. She is also the author or editor of several other books, including *Contagious Capitalism: Globalization and the Politics of Labor in China* (Princeton, 2005). In 2022–2024, Dr. Gallagher will be a Fulbright Global Scholar on a new research project that examines how economic engagement with China has affected domestic public opinion toward globalization. In addition to her academic research, she has consulted with governments, international organizations, and corporations on China's domestic politics, labor and workplace conditions, and urbanization policies. She received her Ph.D. in politics in 2001 from Princeton University and her B.A. from Smith College in 1991.

JENNY J. LEE

Jenny J. Lee is a professor of higher education and Dean's Fellow for Internationalization at the University of Arizona. She is co-editor of the book series *Palgrave Studies in Global Higher Education*. She formerly served as a NAFSA: Association of International Educators senior fellow, U.S. Fulbright Scholar to South Africa, and the chair for the Council of International Higher Education and Board of Directors for the Association for the Study of Higher Education. She is currently the vice president elect for the Postsecondary Education Division of the American Educational Research Association. Her research focuses on the internationalization of higher education, based on her comparative research in the United States, Southern Africa, and East Asia. Her latest research focuses on how geopolitics shape global science, which is covered in her award-winning edited book *U.S. Power in International Higher Education*, published by Rutgers University Press in 2021. She earned her Ph.D. in higher education and organizational change at the University of California, Los Angeles, in 2002.

IVETT A. LEYVA

Ivett A. Leyva became the head of the Department of Aerospace Engineering at Texas A&M University in September 2021. Previously, she was a senior aerospace engineer for the U.S. Air Force for 15 years. She served as the program officer for hypersonic aerodynamics at the Air Force Office of Scientific Research (AFOSR) and as a researcher at the Air Force Research Laboratory (AFRL) working on liquid rocket instabilities. Her technical expertise is in hypersonic aerodynamics and liquid rocket engines. While with the Air Force, she also worked on the protection of basic and applied research. Dr. Leyva holds a bachelor's, master's, and doctoral degree from the California Institute of Technology. She has six patents and has authored numerous papers and two book chapters. She is a fellow of the American Institute for Aeronautics and Astronautics and the Air Force Research Laboratory, and a recipient of a Civilian Achievement Medal and two meritorious Civilian Service Awards and Medals from the Air Force. Dr. Leyva has participated in four National Academies of Sciences, Engineering, and Medicine reports and was a member of the Aeronautics and Space Engineering Board for 6 years.

ELIZABETH D. PELOSO

Elizabeth D. Peloso is the associate vice president and associate vice provost for research services at the University of Pennsylvania (Penn). In addition to managing Penn's federal, other government, and foundation research portfolio, she oversees its compliance program for export controls, and her office is leading Penn's efforts to implement a research security program. Prior to her time at Penn, Ms. Peloso established the University of Delaware's research compliance program, including export controls. She has deep expertise in the management of export control compliance at universities, including the establishment of policy, training, and technology controls as applicable and effective practices. Additionally, she has extensive experience in research contracting and ensuring compliance with regulatory requirements. Ms. Peloso currently serves on the Board of the Council on Governmental Relations, chairing the Research Security and Intellectual Property Committee. She is a past chair of the Association of University Export Control Officers and remains active in the organization. With a background in engineering and business, as well as years of experience in research laboratories, Ms. Peloso brings a practical solution seeking approach to research compliance.

JEFFREY M. RIEDINGER

As vice provost of global affairs at the University of Washington (2013 to present), Jeffrey M. Riedinger has leadership and administrative responsibility for the university's diverse global programming, including support for international research, study abroad, student and faculty exchanges, and overseas centers. As part of his responsibilities, Dr. Riedinger served as director of the Confucius Institute of the State of Washington from 2014 to 2020. He also serves on the faculty of the University of Washington School of Law. He previously served as dean of international studies and programs (2005–2013) and faculty member (1990–2013) at Michigan State University. An expert on the political economy of land reform, Dr. Riedinger has carried out research in East, South, and Southeast Asia, sub-Saharan Africa, Central America, and the Middle East. From 1996 to 2018, he was involved in research on rural land rights in China in collaboration with colleagues at Landesa, a U.S.-based nonprofit, and at Renmin University of China. Dr. Riedinger has conducted briefings on foreign aid, land reform, and international education issues for members of the White House staff, Department of State and U.S. Agency for International Development personnel, and members of Congress. His publications include two books and numerous articles, chapters, reviews, and monographs. He is immediate past president of NAFSA: Association of International Educators. Dr. Riedinger earned a B.A. from Dartmouth College, a J.D. from the University of Washington, and a Ph.D. in public and international affairs from Princeton University.

C. REYNOLD VERRET

C. Reynold Verret is the sixth president of Xavier University of Louisiana. Prior to his 2015 investiture as president, Dr. Verret served as provost and chief academic officer at Savannah State University. Previously, he also served as provost at Wilkes University in Pennsylvania and as dean and professor of chemistry and biochemistry at University of the Sciences in Philadelphia. Dr. Verret also served on the faculty at Tulane University and Clark Atlanta University. His research interests have included the cytotoxicity of immune cells, biosensors, and biomarkers. He has published in the fields of biochemistry and immunology, and also collaborated on matters of social exclusion and health. Throughout, Dr. Verret has worked to increase the number of U.S. students pursuing degrees in science, technology, engineering, and mathematics disciplines and continuing to advanced study. This includes development of qualified science and math teachers in K–12. Dr. Verret received his undergraduate degree cum laude in biochemistry from Columbia University and his Ph.D. in biochemistry from the Massachusetts Institute of Technology (MIT). To these, he added postdoctoral experiences as a fellow at the Howard Hughes Institute for Immunology at Yale University and the Center for Cancer Research at MIT.

Appendix B

Listing of Open, Closing, and Paused U.S. Confucius Institutes

OPEN CONFUCIUS INSTITUTES AT U.S. INSTITUTIONS OF HIGHER EDUCATION

- Alfred University; Alfred, New York
- Pacific Lutheran University; Tacoma, Washington
- San Diego Global Knowledge University; San Diego, California
- Troy University; Troy, Alabama
- Webster University; St. Louis, Missouri
- Wesleyan College; Macon, Georgia

This list does not include Confucius Institutes (CIs) hosted by municipal school districts or other organizations.[1] National Academies of Sciences, Engineering, and Medicine staff contacted institutions of higher education via email and phone to verify the status of CIs.

CLOSING OR PAUSED CONFUCIUS INSTITUTES AT U.S. INSTITUTIONS OF HIGHER EDUCATION

- University of Utah; Salt Lake City, Utah (to close in June 2023)

---

[1] See the National Association of Scholars' list of CIs for further information, available at https://www.nas.org/blogs/article/how_many_confucius_institutes_are_in_the_united_states.

Appendix C

Overview of DOD-Sponsored Fundamental Research

A basic tenet of academic research has long been that fundamental basic science research plays a critical role in maintaining U.S. economic competitiveness and national security. As articulated in National Security Decision Directive 189 (NSDD-189) issued in 1985,[1] and reaffirmed most recently in a memorandum issued by Under Secretary of Defense Ashton B. Carter in 2010,[2] fundamental research is defined as "basic and applied research in science and engineering, the results of which ordinarily are published and shared broadly within the scientific community, as distinguished from proprietary research and from industrial development, design, production, and product utilization, the results of which ordinarily are restricted for proprietary or national security reasons." NSDD-189 further states that the "products of fundamental research are to remain unrestricted to the maximum extent possible. When control is necessary for national security reasons, classification is the only appropriate mechanism."

Introducing the concept of controlled unclassified information (CUI) has blurred the lines for fundamental research (Obama, 2010). The categories of CUI are broad and have created new challenges. A 2019 report to the National Science Foundation (NSF) discussed both the value of fundamental research to the United States and the impact of the introduction of CUI (JASON, 2019). The report concluded that "NSF should support reaffirmation of the principles of NSDD-189, which make clear that fundamental research should remain unrestricted to the fullest extent possible, and should discourage the use of new CUI definitions as a mechanism to erect intermediate-level boundaries around fundamental research

---

[1] Available at https://irp.fas.org/offdocs/nsdd/nsdd-189.htm.
[2] Available at https://fas.org/irp/doddir/dod/research.pdf.

areas." This recommendation might apply equally to basic research funded by the Department of Defense (DOD), which has since created contractual mechanisms to ensure that it does not treat projects including CUI as fundamental research.

THE ROLE OF INSTITUTIONAL RESEARCH SECURITY PROGRAMS

National Security Presidential Memorandum – 33 (NSPM-33), issued on January 14, 2021,[3] directs federal agencies to strengthen the protection of U.S. government-funded research from foreign influence and exploitation. The Office of Science and Technology Policy subsequently issued NSPM-33 implementation guidance to federal agencies in January 2022 (Subcommittee on Research Security Joint Committee on the Research Environment, 2022). This document includes processes for enhanced transparency in disclosure similar to those discussed above, along with the use of persistent digital identifiers. The implementation guidance requires institutions with $50 million dollars or more in annual federal research expenditures to establish formal research security programs to include cybersecurity measures and controls, foreign travel security, and research security and export control training, among other components (Trump, 2021).

The security requirements NSPM-33 imposes serve as a baseline, and additional protections may be required for research that includes generation or use of CUI, such as export-controlled research or research with restrictions on publication or participation. In these cases, the research institution would need to provide enhanced physical security and cybersecurity. For example, this work might require full compliance with NIST 800-171, Protecting Controlled Unclassified Information in Nonfederal Systems and Organizations,[4] as well as physical security plans the academic institution monitors centrally (Ross et al., 2020). Classified research is subject to detailed security requirements outlined in the National Industrial Security Program Operating Manual.

---

[3] Available at https://irp.fas.org/offdocs/nsdd/nsdd-189.htm.
[4] Available at https://csrc.nist.gov/publications/detail/sp/800-171/rev-2/final.

Appendix D

Closure Reasons for U.S. CIs Using National Association of Scholars Data

A spreadsheet of reasons cited for the closure of U.S. Confucius Institutes (CIs) was compiled by National Academies of Sciences, Engineering, and Medicine staff using data collected by the National Association of Scholars.[1] This spreadsheet is not exhaustive, in that it does not contain a listing of every U.S. CI that existed but rather a set of U.S. CIs for which the National Association of Scholars collected closure documents or contract agreements/Memorandums of Understandings. Reasons given for closure, when publicly specified, were further researched and verified by National Academies staff to the extent possible.

---

[1] See https://www.nas.org/reports/after-confucius-institutes/full-report#WhyConfuciusInstitutes Close and https://drive.google.com/drive/folders/1ozgY69PokmXJMWWO-uBy0DaJ021804iv?usp =sharing.

Institution of Higher Education (IHE)	IHE's response to U.S. public policy	Geopolitical relationship with Chinese government	State or IHE budgetary reasons	Lack of interest/ decreased activity/ declining enrollment	CI not aligned with university values/goals	Other reason (no public reason, restructuring, etc.)
Alabama A&M University	X					
Arizona State University	X		X			
Auburn University at Montgomery		X				
Augusta University				X	X	
Baruch College						Possible restructuring
Binghamton University						
Broward County Public Schools						
Bryant University		X				Possible restructuring
California State University, Long Beach	X					

Central Connecticut State University	X		"On July 16, 2020, CCSU was notified that Confucius Institute Headquarters (Hanban) was closing and the CIs would be administered by the Chinese International Education Foundation. With all of the uncertainties these organizational changes bring during these challenging times, CCSU has decided to close its CI on June 30, 2021." – Letter from President Zulma Torro to Hao Pan, CLEC, November 6, 2020
Chicago Public Schools		X	
Clark County School District			"The main reason was we were not able to get licensable teachers to teach Chinese in Nevada." – Shannon La Neve, Director of Humanities, Curriculum and Instruction Division, Clark County School District, via phone to Flora Yan, National Association of Scholars
Cleveland State University		X	Possible restructuring
College of William and Mary	X		
Colorado State University			
Columbia University		X	
Community College of Denver			

continued

Institution of Higher Education (IHE)	IHE's response to U.S. public policy	Geopolitical relationship with Chinese government	State or IHE budgetary reasons	Lack of interest/ decreased activity/ declining enrollment	CI not aligned with university values/goals	Other reason (no public reason, restructuring, etc.)
Dickinson State University				X		
Emory University						
George Mason University						
Georgia State University						"We will support the study of Chinese language and culture through the Office of International Initiatives in cooperation with our valued partner, BLCU. Given the strong and multi faceted links we have developed over the years, including academic collaborations, student exchange and study abroad, and its status as one of the leading institutions for teaching Chinese as a foreign language, BLCU is the ideal partner for us to build on and expand the many achievements of the Georgia State CI. The staff formerly affiliated with the CI will transfer to this new initiative and continue to support teaching and outreach efforts at the university and in the Atlanta community." – Email from Associate Provost for International Initiatives Wolfgang Schlör to undisclosed recipients, August 19, 2020

Institution				Reason
Houston Independent School District				
Indiana University–Purdue University	X			Possible restructuring
Kansas State University				Possible restructuring
Kennesaw State University		X		
Medgar Evers College				
Miami Dade College			X	
Miami University (Ohio)				Possible restructuring
Michigan State University	X			Possible restructuring
Middle Tennessee State University				Possible restructuring
New Jersey City University				No stated reason
New Mexico State University			X	Possible restructuring
North Carolina State University				Possible restructuring

continued

Institution of Higher Education (IHE)	IHE's response to U.S. public policy	Geopolitical relationship with Chinese government	State or IHE budgetary reasons	Lack of interest/ decreased activity/ declining enrollment	CI not aligned with university values/goals	Other reason (no public reason, restructuring, etc.)
Northern State University					X	Possible restructuring
Northwest Nazarene University	X	X				
Old Dominion University	X				X	
Pace University						Possible restructuring
Pennsylvania State University		X			X	
Pfeiffer University						"The CI that formerly was located at Pfeiffer moved to the University of North Carolina at Charlotte at that time." – Email from President Colleen Keith to the Wilberforce Institute, June 2018
Portland State University	X		X			Possible restructuring
Prairie View A&M University	X					
Presbyterian College				X		COVID-19 cited
Rutgers University	X		X			

San Diego State University			"SDSU closed and transferred the [Confucius Institute] on June 30, 2019 to San Diego Global Knowledge University. "Instead, the university is "launching a new Chinese Cultural Center." – In "New Chinese, Global Education Center Launched at SDSU," SDSU News Center, August 7, 2019, https://newscenter.sdsu.edu/sdsu_newscenter/news_story.aspx?sid=77714
San Francisco State University	X		
Savannah State University			"Current world circumstances with the force majeure presented by the global COVID-19 pandemic dictate that we discontinue the collaboration, and close the Confucius Institute at Savannah State University." – Letter from Interim President Kimberly Ballard Washington to Ma Jianfei, CLEC, July 1, 2020
Southern Utah University			
St. Cloud State University		X	
Stanford University			
State College of Optometry, State University of New York			
State University of New York at Albany	X		Possible restructuring

continued

Institution of Higher Education (IHE)	IHE's response to U.S. public policy	Geopolitical relationship with Chinese government	State or IHE budgetary reasons	Lack of interest/ decreased activity/ declining enrollment	CI not aligned with university values/goals	Other reason (no public reason, restructuring, etc.)
State University of New York Global Center						
Stony Brook University						Possible restructuring
Temple University						Possible restructuring
Texas A&M University	X					Possible restructuring
Texas Southern University		X				
The George Washington University						
Tufts University						"We have decided to focus more on our strong and growing direct relationship with BNU." – James M. Glaser (Dean of the School of Arts and Sciences) and Diana Chigas (Senior International Officer and Associate Provost) in "Decision to Close the Confucius Institute at Tufts University," Office of the Provost and Senior Vice President, Tufts University, March 17, 2021, https://provost.tufts.edu/blog/news/2021/03/17/decision-to-close-the-confucius-institute-at-tufts-university/

Tulane University			No stated reason
University at Buffalo		X	Possible restructuring
University of Akron	X		
University of Alaska Anchorage		X	
University of Arizona	X		
University of California, Davis		X	
University of California, Los Angeles		X	COVID-19 cited
University of California, Santa Barbara			No stated reason
University of Central Arkansas	X		
University of Chicago		X	No statement
University of Delaware		X	"It was simply a decision made due to decreased activity." – Associate Deputy Provost Ravi Ammigan, quoted in Kirk Smith, "Confucius Institute Set to Close in Early 2020," *The Review*, October 8, 2019, http://udreview.com/confucius-institute-set-to-close-inearly-2020/

continued

Institution of Higher Education (IHE)	IHE's response to U.S. public policy	Geopolitical relationship with Chinese government	State or IHE budgetary reasons	Lack of interest/ decreased activity/ declining enrollment	CI not aligned with university values/goals	Other reason (no public reason, restructuring, etc.)
University of Hawai'i Mānoa	X					
University of Idaho						Possible restructuring
University of Illinois at Urbana-Champaign			X			
University of Iowa			X			
University of Kansas	X					Possible restructuring
University of Kentucky	X					
University of Maryland	X					
University of Massachusetts Boston					X	Possible restructuring
University of Memphis					X	
University of Michigan						Possible restructuring
University of Minnesota	X					

University of Missouri		X		
University of Montana			X	
University of Nebraska–Lincoln		X		Possible restructuring; COVID-19
University of New Hampshire	X			Possible restructuring
University of North Carolina at Charlotte	X			Possible restructuring
University of North Florida			X	
University of Oklahoma	X	X		
University of Oregon	X			Possible restructuring
University of Pittsburgh	X			
University of Rhode Island	X			
University of South Carolina			X	
University of South Florida	X			

continued

Institution of Higher Education (IHE)	IHE's response to U.S. public policy	Geopolitical relationship with Chinese government	State or IHE budgetary reasons	Lack of interest/ decreased activity/ declining enrollment	CI not aligned with university values/goals	Other reason (no public reason, restructuring, etc.)
University of Southern Maine		X		X		
University of Tennessee, Knoxville					X	Possible restructuring
University of Texas at Dallas					X	Possible restructuring
University of Texas at San Antonio						Possible restructuring
University of Toledo						
University of Washington						Transferred CI to Pacific Lutheran University
University of West Florida				X		
University of Wisconsin–Platteville	X	X			X	
Valparaiso University	X		X			
Wayne State University						Possible restructuring

West Virginia University			X	
Western Kentucky University	X			Possible restructuring
Western Michigan University		X	X	COVID-19 cited
Xavier University of Louisiana				"The University continues to offer for-credit Chinese language courses and the Chinese Minor through its Department of Languages." – "Confucius Institute," Xavier University, https://web.archive.org/web/20220225051341/https://www.xula.edu/confucius/
TOTAL	**31**	**9**	**16**	**9**	**20**

Appendix E

Information-Gathering Sessions with U.S. Colleges and Universities That Are Current or Former Hosts of Confucius Institutes

A Report to the National Academies of Sciences, Engineering, and Medicine Committee on Confucius Institutes at U.S. Institutions of Higher Education

Korantema Kaleem, Jasmine Howard, and Kellie Macdonald

American Institutes for Research

OVERVIEW

The National Academies of Sciences, Engineering, and Medicine (the National Academies) Confucius Institutes Study Committee was tasked with conducting a consensus study to (a) identify best practices and principles regarding appropriate operations for U.S. academic institutions and (b) share information that the Department of Defense (DOD) could use regarding the potential issuance of waivers for the prohibition of research support at institutions that host Confucius Institutes (CIs).

As part of this consensus study of institutions hosting CIs, the National Academies contracted with the American Institutes for Research (AIR) to conduct information-gathering sessions with U.S. colleges and universities (institutions) that are currently hosting, or have hosted, CIs and that participated in the 2019 National Defense Authorization Act (NDAA) waiver process. In summer 2022, the National Academies staff compiled a list of target institutions for the information-gathering study sample, including contact information for potential respondents from each institution. The AIR research team collaborated with the National Academies in the outreach and recruitment of the target institutions.

These recruitment efforts resulted in a final sample of seven institutions. Across the seven institutions, AIR completed virtual information-gathering sessions with 10 individuals identified as having knowledge of and familiarity with their institution's CI.[1] The focus of the conversations was on institutions' experiences and perspectives related to CI operations and the 2019 DOD waiver process.

In July 2022, AIR presented preliminary findings from the information-gathering sessions to the committee. This report provides a more detailed account of the final set of findings and is intended to inform the broader National Academies consensus study. The results will be synthesized in two reports authored by the committee. Findings from these two consensus study reports, including the findings discussed in this report, may be used to inform federal processes regarding waiver criteria and the conditions under which a waiver may be considered and granted to allow an institution to host a CI and receive federal agency funding.

DATA COLLECTION AND ANALYSIS

This report presents the key findings from AIR's information-gathering sessions with seven institutions to answer the following research questions (RQs):

- RQ1. How are institutions with CIs operated and funded?
- RQ2. What are the characteristics of CIs and their partnerships with Hanban?[2]
- RQ3. What are or were the most important factors that led to the closure of CIs?
- RQ4. What measures do or did institutions have in place to preserve academic freedom, openness, and/or national security in 2019?
- RQ5. What were the experiences of institutions' staff who applied for DOD waivers in 2019?

To address these research questions, AIR conducted 60-minute, virtual, information-gathering sessions with 10 senior-level CI administrators across a sample of seven institutions representing both private and public colleges and universities. The information-gathering sessions were conducted over a 3-week period between June 2022 and July 2022 through a semistructured interview protocol.[3]

---

[1] One respondent was a senior-level administrator at a CI at a K–12 public school.

[2] "Hanban" is the colloquial term for the Chinese International Education Foundation, or CIEF, now known as the Ministry of Education Center for Language Education and Cooperation, or CLEC. This is the Chinese government agency affiliated with China's Ministry of Education that promoted, managed, and funded CIs on foreign campuses.

[3] A semistructured interview protocol allows for flexibility when obtaining qualitative data. It provides structure with a list of open-ended questions all respondents are asked and permits individual follow-up questions, or probes, to mine for deeper or more detailed information from a respondent.

APPENDIX E

Protocol questions addressed respondent experiences with hosting a CI related to operations, partnerships, CI closure decisions and processes, risk-mitigating measures, and the 2019 DOD waiver process. AIR developed and finalized the protocol with direct input, consultation, and collaboration with National Academies staff. The full information-gathering protocol is provided at the end of this report.

Institution and Respondent Characteristics of Study Sample

The seven institutions that participated in the information-gathering sessions are in different locations in the United States. Of the seven institutions, four are public and three are private. Five of the seven institutions represented in the sample were open for more than a decade. Four of the seven CIs had a specific linguistic and cultural focus, and the other three CIs focused on promoting a general appreciation of Chinese language and culture. Three institutions are open, and four have closed; all are 4-year institutions (see Table E-1).

TABLE E-1 Institution and Respondent Characteristics

Public/Private	Undergraduate enrollment	CI closed or open	Applied for a DOD waiver in 2019	DOD waiver received in 2019	Number of years as CI administrator
Private – 4 years	<10,000	Open	Yes*	No	4 years
Public – 4 years	>10,000	Closed	Yes	No	12 years
Public – 4 years	>10,000	Closed	Yes	No	13 years
Private – 4 years	<10,000	Open	No	No	3 years** 2 years**
Public – 4 years	>10,000	Closed	No	No	7 years
Public – 4 years	>10,000	Closed	Yes	No	9 years** 7 years**
Private – 4 years	<10,000	Open	No	No	8 years

NOTES: CI is Confucius Institute. DOD is Department of Defense.
* Indicates that an institution applied for a DOD waiver, but the respondent was not involved in the waiver application because it was handled by a separate department.
** Indicates that two administrators from the same institution were interviewed.
SOURCE: U.S. News & World Report, 2022.

DATA ANALYSIS

The virtual information-gathering sessions were audio-recorded and transcribed using the Zoom meeting platform. The transcripts were uploaded into the NVivo qualitative software data analysis program and systematically coded by two AIR researchers using a codebook that aligned with the key research questions (RQ1–RQ5, above) and more specific focal areas of interest identified by National Academies staff. The full set of information-gathering questions is provided at the end of this report.

To ensure intercoder reliability, each analyst coded two transcripts separately at the outset of the coding process to gauge consistency in code application and coding agreement. The code definitions were adjusted in the codebook as needed. After the team reached the intercoder agreement threshold of 95 percent, the codebook was finalized and data analysis ensued. The AIR team engaged in regular communication to monitor and ensure consistency throughout the coding process.

Limitations of the Data

The data and findings presented in this report should be interpreted with two key considerations: (a) The study timeline necessitated a short window of time to recruit, schedule, and conduct information-gathering sessions with institutions that hosted a CI. The timeline affected the ability of some institutions to participate due to the conflicting schedules of respondents or, in some cases, competing demands on their time (including other projects related to CIs with timelines overlapping this study timeline); and (b) staff turnover affected some institutions' ability to participate when administrators with experience and familiarity with an institution's CI were no longer employed at the institution. These challenges to recruitment and participation resulted in a smaller sample of institutions than initially targeted and reflect the experiences of a relatively small number of institutions—7 out of more than 100 institutions that have had or currently still host CIs in the United States (Horsley, 2021; NAS, 2022).

FINDINGS

The findings discussed in this section are based on the virtual information sessions. Respondents from the seven participating institutions had their own unique insights about their experiences hosting a CI, and a few provided insights about their experiences with the DOD waiver process.

The sections below highlight key findings relative to the five research questions, which represent areas of interest to the committee.

CI Operations and Funding

All respondents were asked to elaborate about their institution's CI operations and funding structure. The following section highlights findings related to CI leadership and staffing, CI affiliations with academic departments, CI funding and budget structure, and CI programming.

CI Leadership and Staffing

Leadership Structure. Institutions included in this study maintained a leadership structure consisting of at least one director from the U.S. institutions and a leadership board of some kind (e.g., board of directors, board of advisors). A few CIs were led by co-directors; typically, the co-director was an administrator from the Chinese partner institution.

All respondents indicated that their CI was guided by an advisory board or a board of directors. These boards typically were composed of senior-level administrators from both the U.S. and Chinese partner institutions. Respondents reported that board members held a variety of positions at their host institution, including president of the institution, provost, dean, and faculty. Board involvement and influence in CI operations varied across institutions. One respondent said the board of their CI had minimal involvement in CI operations, and that over a 4-year period the board met once a year to receive an overview of operations and events at the CI. In this case, the CI director was solely responsible for approval of programming and expenditures. Other boards were described as being more involved in CI operations. One respondent explained that they had to receive approval from their board before submitting budget proposals to Hanban, although most programming and operations decisions were made by CI directors. The instructor staffing process included the following components: applicant recruitment, applicant screening, and application selection.

Staffing. Respondents from all seven institutions described the processes used to select CI instructors, including screening and recruitment.

The process began with the Chinese partner institutions issuing a public call for applications to work at a specific U.S.-based CI. To narrow down the pool of applicants, Hanban and the Chinese partner institutions conducted their own screening process, including personal interviews and language and culture tests, to ensure applicants were well qualified to work at a U.S. institution.

Some institutions understood that their location could serve as either an incentive or a disincentive for attracting, hiring, and retaining instructors. In other words, applicants were drawn to an institution's location for a variety of reasons, including the climate and the geographic location of the institution. For some institutions, location poses a challenge to recruiting and retaining applicants. In particular, one respondent said their location in a rural setting serves as a deterrent, and they sometimes find it difficult to attract instructors.

Once the Chinese partner institutions narrowed down the pool of applicants, the U.S. institution assumed control of the hiring process, which included a review of applicant résumés/curricula vitae and then a series of in-person or virtual interviews using their own set of criteria for hiring CI instructors. One respondent explained that their CI followed the same criteria for hiring instructors as the academic department in which the CI was housed. The full information-gathering protocol is provided at the end of this report.

Respondents across the seven institutions also described the instructor evaluation process as well as issues they experienced with staffing their respective CIs.

Evaluating Instructors. Student evaluations served as the main source for evaluating the performance of CI instructors across the institutions, with most respondents indicating that this was generally consistent with their institution's practice for evaluating faculty and instructors. However, according to one respondent, the CI director and the head of the department in which the CI was housed at the institution reviewed and assessed CI instructors and staffing needs each term.

Issues with Staffing. Some respondents described having very few issues related to CI instructors, while others reported that their CI did not have any issues. Those who reported issues described them as minor and related mainly to logistics (e.g., housing, finances) or to the Chinese instructors experiencing challenges with adjusting to the life and culture of their host institution. While some of these issues were resolved through instructional coaching and mentoring, in a few cases, CI instructors' contracts were terminated prematurely. Decisions to end a contract early were described as the responsibility of the CI director.

CI Affiliation with Academic Departments

All respondents were asked to describe how their CI was affiliated with academic departments and the physical location of the CI at their institution. Some CIs included in the study were affiliated with academic departments; however, the structure of these affiliations varied by institution, and CI directors typically held other positions or roles beyond their position as the CI director (i.e., not a designated full-time position). For example, the CI directors often held faculty appointments or additional leadership roles within their academic departments.

Department Affiliation. Most respondents indicated that their CI fell under the Department of East Asian Studies, foreign language department, or comparable academic unit. Others, however, were housed across departments affiliated with international studies or languages but not directly affiliated with a specific department. In one case, the CI was housed under the president's special projects.

Regardless of where the CI was officially housed, respondents across the seven institutions noted that collective academic events such as teacher immersion trainings, seminars, and other events included the CI and affiliated departments.

At six of the seven institutions, major degree-granting programs did not fall under the CI directly but, rather, across various related departments. The degree offerings ranged from undergraduate to advanced professional degrees, with CI-specific courses counting toward degree credit. However, not all CI courses across the seven institutions were credit bearing.

Physical Location. All respondents reported that the CI at their institution had a physical location on their campus. Some CIs had a small, designated space, whereas CIs at other institutions were located in an area accessible by the campus community (students, faculty, staff, and visitors). Some CIs shared space with other academic departments.

CI Funding and Budget Structure

CI funding structures fell into one of three types: (a) institutions received funding from Hanban or a Chinese third-party agency and its host institution, (b) funding from Hanban or a Chinese third-party agency and its host institution typically matched the funds, and (c) Hanban or a Chinese third-party agency provided funds, whereas U.S. host institutions provided in-kind support (i.e., office, event space, furnishings, or classrooms). Some institutions received funding up front, while others received an initial gift and annual funding. Others received annual funding from Hanban or a Chinese third-party agency. Overall, across the seven institutions, respondents reported that the funding and in-kind support allocated to CIs enabled institutions to continue daily operations. Most respondents also discussed their budget structure and how funds were allocated.

Funding and In-Kind Support. Several respondents described the funding split as "50/50" between Chinese and American sources. Most of the funding from Chinese sources supported cultural programming offered by the CIs, with the remainder used to fund operating expenses, Chinese staff salaries, and trainings for staff. Alternatively, funding from U.S. sources was described as "in kind" because it was allocated to supporting the physical space occupied by the CI and the salaries of staff employed by the U.S. institution that supported the CI. One respondent explained that additional funds were available from Hanban and did not come out of the annual budget. These additional funds included up to $3,000 for additional materials as well as travel expenses for attending conferences or convenings in China. A few CIs received funding from other sources, including the local community, fundraising efforts, and online learning programs.

To secure funding from Hanban, several respondents explained that CI directors submitted annual budget proposals to Hanban outlining planned activities for the year as well as estimated costs of these activities. Within the annual budget proposals, CIs requested specific dollar amounts to fund their proposed activities for the year. Respondents reported budget proposals ranging from $40,000 to $400,000. For some CIs, funding requests were consistent from year to year, whereas for others, funding requests varied (i.e., typically increased) over time. Though several respondents indicated that they rarely were denied funding for the activities outlined in their budget proposals, respondents noted that Hanban made the final decision as to whether proposed activities would receive funding. Most respondents were unaware of whether their CI received any one-time gifts, except for one respondent who reported that their CI received startup funds in addition to their yearly installment.

Budget Structure. Most CIs allocated the majority of their budget for cultural programming (e.g., music performances, calligraphy workshops). One respondent explained that greater than 80 percent of their budget was spent on cultural programming (e.g., cultural events, talks by visiting scholars, musical performances). Other CIs allocated more of their budget toward organization and delivery of courses, including salaries for instructors, trainings, and professional development. In one instance, a CI budgeted for Ph.D. student fellowships. One respondent explained that funds were also allocated for miscellaneous expenses (e.g., supplies, technology, faculty reimbursements).

One respondent indicated that the CI budget is entirely separate from the institution's budget, and that funding from Hanban never became part of the institution's operational fund. When asked about the use of Hanban funds, all respondents reported that their CI never used funding from Hanban to support scientific research. Rather, all funding was used to support language instruction and cultural programming.

CI Programming

CI programming varied by institution. Some CIs offered cultural programming only (e.g., Chinese New Year celebrations), whereas others offered Chinese language classes only. While some language classes were offered for credit, other CIs only offered not-for-credit classes.

Regardless of whether a CI offered for-credit or not-for-credit classes, cultural programming activities and classes at each CI across the seven institutions were open to students at the institution and members of the surrounding community. Respondents described their CI programming in detail. The programming included curriculum development for CIs that delivered Chinese language classes and activities beyond instruction. One participant explained that it would be impossible to foster learning and appreciation related to the Chinese culture without involving Chinese entities.

Curriculum Development. For the CIs that delivered Chinese language classes, the responsibility for developing the curriculum varied by CI; however, several respondents stated unequivocally that Hanban did not have any say in the curriculum delivered by the CI. For example, one respondent explained that they never received any directive from Hanban on what they had to teach at the CI, and another explained that the curriculum development process was driven entirely by the U.S. host institution. A few CIs included text in their agreements with Hanban stating that the U.S. institution ultimately approved the CI curriculum.

A few respondents reported that the curriculum was developed by the faculty teaching the courses. One respondent explained that their CI had an understanding with Hanban that oversight related to curriculum delivery was required to ensure that all instructors adhered to the same standards. At other CIs, the curriculum was established by the academic department under which the courses were housed. In those instances in which the curriculum was set by the academic department, the CI provided an instructor to deliver the predetermined curriculum. One respondent stated, however, that at their CI, the instructors from China were responsible for developing the CI curriculum.

Three institutions reported that all CI curricula underwent their institution's rigorous review process, just like any other course being taught at the institution. One respondent noted that at their CI, a third party reviewed all CI curricula to ensure that they met the institution's standards. In terms of materials, just one respondent reported that their institution uses textbooks recommended by Hanban; however, these books still go through the institution's textbook-vetting process.

Activities Beyond Instruction. All CIs across the seven institutions offered activities other than class instruction. These activities included procuring items for local museum exhibits related to China, planning summer trips to China for local high school students, organizing and facilitating calligraphy workshops, hosting conferences, and providing cultural programming at local K–12 schools. In addition, several CIs were approached by their local community or by local performers to co-sponsor events such as festivals, music performances, poetry readings, and dance recitals. The responsibilities tied to co-sponsorship of these activities varied by CI, with CIs providing a combination of funding, space, staffing, or programming to support the activities in their local community. Few respondents were able to speak to the approval process for these community-focused events; those who did reported no issues with gaining approval for activities outside of instruction.

Relationships with Hanban and Chinese Partner Institutions

For the purposes of this report, a formal relationship is defined as a legal contract with two or more entities, and an informal relationship is without a legal contract. Across the seven institutions participating in this data collection, the relationships between the U.S. host institution and the Chinese partner varied in formality but were typically initiated by the U.S. institutions. Several respondents reported that the relationship with their Chinese partner was initiated by a member of the faculty and/or an administrator based on an existing relationship; however, in at least one case, the relationship with the Chinese partner began primarily with the establishment of the CI. The AIR research team noted that two institutions participating in information-gathering sessions have the same Chinese partner institution.

Pre-Existing Relationships with Chinese Partners

Several respondents described having informal relationships with their Chinese partner institution, outside of or in addition to their official CI agreement. One participant stated that the relationship with their Chinese partner was established due to an alumnus of the university visiting the partner institution. Through this individual a connection was formed, and the relationship between the U.S. institution and the Chinese partner was established. Another respondent stated that their faculty established relationships within their CI's Chinese partner institution by conducting research in China.

Formal Agreements to Host a CI

Institutions with formal partnerships described their institution's CI arrangement as a contracted partnership between their institution, Hanban, and a Chinese partner university. In some cases, there was one agreement with the partner institution and Hanban; however, at least one institution negotiated separate formal contracts with the Chinese partner institution and Hanban. Although the details of these institutions' formal agreements varied, they generally shared a common purpose and goal: to cultivate appreciation for Chinese culture within the institution as well as in the local community.

Supplemental MOUs. Most respondents reported that their institution did not sign a supplemental Memorandum of Understanding (MOU) with their Chinese partner institution; however, a few respondents said their institution signed a supplemental MOU with their Chinese partner. Supplemental agreements signed with a Chinese partner institution outlined how the governing board would form and the frequency with which they would meet. A respondent indicated that the terms of the supplemental agreement restricted the authority of the governing board, which serves in an advisory role.

CI CLOSURE

The announcement by DOD in 2019 to deny funding to institutions of higher education that host a CI, unless those institutions applied for a 2019 NDAA waiver, caused institutions to make abrupt decisions regarding the future of their

APPENDIX E

CI (CRS, 2022; GAO, 2019). The announcement in 2019 by the government served as the deciding factor for some institutions to close their CI. One respondent explained that the decision to close their CI came directly from the university president; another respondent said the decision came from their university leadership.

Of the four institutions in the sample whose CIs closed after the announcement, three institutions' programs were integrated into other departments (e.g., Chinese language courses), and other programming was taken up by local organizations to ensure continuation. One participant explained that when their CI closed, the language instruction reverted to its "pre-Confucius Institute state," during which it was housed under an academic department. Few respondents said they still have an informal relationship with their partner Chinese institution after the closure of the CI, with some describing enduring friendships with Chinese staff post-CI closure.

Participation in the 2019 NDAA Waiver Process

Institutions that applied for the DOD waiver were asked about their experiences with the 2019 NDAA waiver process. Some respondents who went through the process provided information about their institution's decision to apply for a waiver or close their CI and about their experiences with the application process.

Decision to Apply for a Waiver or Close CI

Four institutions included in this study decided to move forward with the 2019 NDAA waiver process; however, of the respondents included in the information-gathering sessions, three institutions had direct experience with the process. One respondent of a CI that is currently open reported that the waiver application was handled in another department; therefore, they could not provide any information or perspectives on the details of the application process. Three institutions, now closed, indicated that the decision to apply for a waiver was contingent on the ways in which the CI brought value to their institution and the surrounding community. In one case, a CI brought value to their institution by helping to create a pipeline for Chinese language instruction, whereby the CI offered introductory language courses and a separate department offered intermediate and advanced courses. Another institution decided to apply for a waiver because they had a strong partnership with the local K–12 school district to provide Chinese language instruction, which would have been dissolved with the closure of the CI at the university. One institution decided to close their CI instead of applying for a waiver because they wanted to ensure that their federal funding was not jeopardized.

Application Process

Two institutions that hosted CIs that are now closed had experience with the process. They stated that they applied because they felt confident that they would be granted a waiver because their CI had separate leadership and the academic departments received DOD funding. Of these two institutions, one explained that their university leadership considered their CI to be a good candidate for receiving a waiver because the CI funding was separate from DOD funding. As part of the waiver process, these institutions provided the DOD with documents from different departments within the institution. Required documents included institutions' signed agreements with Hanban and annual reports as well as other documents from all of the academic departments within the institution. One respondent explained that their institution gathered every agreement that each academic department had with China to submit along with their application materials.

Three respondents whose institutions applied for a waiver and are now closed expressed disappointment and frustration with the process, citing a lack of communication from the DOD upon submission of the application as well as informal hints along the way that waivers would not be granted. As a result of these types of experiences during the process, one institution decided to formally withdraw their application. Ultimately, these three institutions that applied for a waiver were denied. None of the applicants received an explanation for the denial.

Campus Precautions and Risk-Mitigating Measures

The following sections highlight findings on the respondents' perceptions of risk; the steps that institutions took to mitigate risk and preserve academic freedom, openness, and national security; and the steps that institutions took in evaluating partnerships involving funding in other countries. Of the seven institutions represented in the study, three institutions reported that their university conducted classified research. These institutions clarified that classified research was conducted in other departments at the institution and their CIs did not conduct classified research. In addition, more than half of the respondents stated that their institution works with classified or controlled unclassified information, and the remaining respondents said their institution does not work with this information or could not answer the question.

Perceptions of Risk, 2019 to the Present

Several respondents said they were aware that local and national politicians scrutinized the CIs. Some respondents explained that the scrutiny had increased because politicians questioned the motives of the Chinese government and their involvement in CIs and in U.S. institutions of higher education, and politicians expressed concern about the potential threat that CIs posed to U.S. national security. Some respondents shared that the Federal Bureau of Investigation (FBI) contacted them for clarification of the role of the Chinese government in their

APPENDIX E

CI. One respondent, who was the director of a CI at a public research institution, shared that before their CI closed, the FBI contacted them to inquire about Chinese involvement in their CI.

All respondents said they believed that their CI did not pose a national security risk. Respondents cited the focus of their CI (on cultural awareness), the lack of scientific research taking place, and the long-standing relationships with Chinese staff as reasons that their CI did not constitute a national security risk. Additionally, when asked about preserving academic freedom, openness, and national security, some respondents explained that these issues were not of high concern to them because they had maintained friendships with the Chinese staff, which fostered a sense of openness at the CI. A few respondents commented on academic freedom specifically, with one respondent explaining that Hanban did not restrict "normal operations" of the institution, particularly regarding academic freedom and research. Another respondent explained that the CI preserved academic freedom by using materials (e.g., textbooks) from trusted curriculum providers rather than materials provided by Hanban.

Risk Mitigation Measures

Despite respondent sentiments that CIs do not pose a risk to academic freedom, openness, and national security, respondents mentioned that their institutions followed specific processes—that is, processes not specific to CIs—to ensure that staff, faculty, and administrators were aware of possible threats. Most of the mitigation measures described were internal to the institution; however, some CI leaders and staff participated in annual seminars hosted by the FBI, which provided training to increase awareness of potential threats when working with foreign entities.

Most respondents discussed the involvement of various offices within the institution in mitigating risks associated with a foreign entity such as Hanban. These offices typically offered governance of legal agreements and robust security processes for visa procurement (e.g., instructors were prescreened and vetted and the office verified documents).

One respondent reported that their institution's office of international affairs was involved in the visa procurement process for foreign instructors. This institution systematically conducted an institution-wide audit process of each office every 6 or 7 years across various departments. This audit process served as an additional layer of protection to assess individuals working with minors.

Another respondent described the roles of their institution's office of data governance and legal department, which vetted their agreements with Hanban before the agreements were signed. Two respondents mentioned working with an export control officer. The officer conducts background checks on visiting scholars and ensures that any scholars associated with the university who enter or exit the country are in compliance with U.S. and foreign laws.

As another example, one institution worked with their national security and research review committee, which was described as being in charge of understanding federal regulations related to working with a foreign entity. Similarly, another institution worked with its office of research administration, which is responsible for reviewing research or activities funded by a foreign entity. In addition, this institution's CI had a group of top leaders from the institution who were involved in a committee to monitor agreements and use of funding from foreign sources such as Hanban.

Evaluating Partnerships Involving Foreign Funding

For most of the seven institutions in this study, management of foreign gifts occurs at the university level. Respondents explained that their institutions' respective research offices were responsible for evaluating foreign gifts and that these offices often worked with offices of international affairs as well as offices of financial aid.

The ways in which CIs treated one-time gifts varied by institution. Although some respondents were not clear on their institution's policy regarding one-time gifts, one participant explained that their institution does not have a formal policy. Two respondents said their institution does not treat one-time gifts differently than annual installments, and one of the respondents explained that their institution's process remains the same regardless of the gift amount.

CONCLUSION

Through the virtual information-gathering sessions, the respondents from the seven institutions included in this data collection effort shared their experiences and perspectives on their CI operations, funding structures, activities, and programming; the benefits and challenges related to their CI; and the DOD waiver application process. The primary aim of the data collection and analysis is to assist the Confucius Institutes study committee in understanding the direct experiences of hosting a CI and the DOD waiver process from institutions that applied in 2019.

Overall, respondents cited many benefits of the CI and found the relationship and CI activities beneficial and rewarding to the institution, the students, and the local community. Issues with the Chinese partners, Hanban, and/or Chinese instructors were infrequent and generally described as minor. Commonly shared experiences and perceptions across the seven institutions were as follows: (a) respondents' CIs were described as successfully fostering appreciation of the Chinese culture across their institution and local communities; (b) CIs were described as bringing awareness of Chinese language and culture to U.S. institutions; and (c) although respondents were aware of criticisms and concerns from local and national politicians related to CIs and the role of the Chinese government in higher education, they believed these concerns to be unfounded and unwarranted. Respondents pointed to their agreements with Hanban as well as several other internal measures at their institutions to mitigate risk associated with inappropriate Chinese government involvement.

Given the overall positive experiences with CIs, respondents shared a common disappointment with the concerns raised about U.S. institution-based CIs and expressed frustration with the 2019 DOD waiver application process and the limited communications they received about why waiver applications were denied. Some institutions were able to work with the local community and internally within their academic departments to continue activities and classes they perceived as particularly beneficial after the official closure of their CI.

REFERENCES

CRS (Congressional Research Service). 2022. *Confucius Institutes in the United States: Selected Issues*. Retrieved from https://crsreports.congress.gov/product/pdf/IF/IF11180.

GAO (Government Accountability Office). 2019. *Agreements Establishing Confucius Institutes at U.S. Universities Are Similar, but Institute Operations Vary* (GAO-19-278). Retrieved from https://www.gao.gov/assets/gao-19-278.pdf.

Horsley, J. P. 2021. It's time for a new policy on Confucius Institutes. *Brookings Institution*, April 1, 2021. Retrieved from https://www.brookings.edu/articles/its-time-for-a-new-policy-on-confucius-institutes/.

NAS (National Association of Scholars). 2022. How many Confucius Institutes are in the United States? NAS blog, April 5, 2022. Retrieved from https://www.nas.org/blogs/article/how_many_confucius_institutes_are_in_the_united_states.

U.S. News & World Report. 2022. *Education* (web page). U.S. News Best Colleges. Retrieved from https://www.usnews.com/best-colleges.

INFORMATION-GATHERING QUESTIONS

Background Questions

- What is/was your title and institutional affiliation?
- How long have you been/were you in this role?
- What is your connection to your institution's Confucius Institute?

Campus Confucius Institute Information

1. Now, we would like to begin by learning more about your Confucius Institute.
    - What year did your Confucius Institute open/begin? What year did it close/end, if applicable?

- Does/Did your Confucius Institute have a specific language and culture focus area? If so, what is/was it?
- Who is/was your Chinese partner institution?
2. Did your institution have a formal (contract and/or MOU) or informal relationship with your Chinese partner institution prior to establishing your Confucius Institute?
3. **[Only CIs that have closed]** Following the closure of your Confucius Institute, has your institution maintained a formal (contract and/or MOU) or informal relationship with your Chinese partner institution? If yes, please describe.

Campus Confucius Institute Operations

4. We would like to learn about the agreement that your institution signed with Hanban to host a Confucius Institute.
   - Did your institution negotiate any parts or clauses contained in this agreement?
   - Did your institution sign any sort of supplemental MOU with your Chinese partner institution?
   - Is your agreement with Hanban publicly available? If so, can you point us to a link?
5. Where, both physically and organizationally, if applicable, was your Confucius Institute located relative to the rest of your academic institution?
   - Probe: Was the Confucius Institute physically located on campus or off campus? Please explain.
   - Probe: Was the Confucius Institute affiliated in some way with an academic department or division? Please explain.
6. What was the relationship between your Confucius Institute and the Department of East Asian Studies, Department of Languages, or comparable academic unit at your institution?
   - Does your institution have a major or degree-granting program related to China?
7. What was the leadership structure of your institution's Confucius Institute?
   - What is/was the breakdown (percentage-wise) of U.S. and Chinese staff at your institution's Confucius Institute?
   - Whom does/did your institution's Confucius Institute director report to?
   - Does/Did your institution have a board of directors to oversee the Confucius Institute?

8. Did any of the leaders or staff of your institution's Confucius Institute hold other university affiliations, such as a faculty appointment?
9. How and by whom was your Confucius Institute funded?
   - Did your Confucius Institute receive a large initial gift?
   - Did your Confucius Institute receive annual funding installments from Hanban, regardless of whether you received a large initial gift?
   - Did contracted amounts of funding from Hanban match amounts received by your institution?
   - Did your institution ever use funding from Hanban to support scientific research?
10. Can you share more about your Confucius Institute's yearly budget outlay? Approximately what was the overall yearly budget for your institution's Confucius Institute, and what were the major categories of expenses?
11. How and by whom was the curriculum at your Confucius Institute developed?
    - Did for-credit students at your institution utilize the Confucius Institute's curriculum?
    - Did your institution utilize texts and course materials provided by Hanban?
    - What types of issues, if any, did your institution encounter related to your Confucius Institute's curriculum?
12. How and by whom were the instructors at your Confucius Institute selected and evaluated?
    - What types of issues, if any, did your institution encounter related to your Confucius Institute's instructors?
    - Were any instructors asked to leave? If so, who made the decision to ask the instructor to leave, and how was this done?
13. What other activities, beyond instruction, did your Confucius Institute offer?
14. How and by whom were events sponsored by or at your Confucius Institute approved?
    - Was an external organization such as a local consulate ever involved?
    - What types of issues, if any, did your institution encounter related to your Confucius Institute's events?

Campus Precautions and Risk-Mitigating Measures

15. What general precautions and risk-mitigating measures, if any, did your institution have in place to preserve academic freedom, openness, and/or national security in 2019?
    - Have these changed at all since 2019?
16. What general precautions and risk-mitigating measures, if any, did your institution have in place to evaluate partnerships involving foreign funding and gifts in 2019?
    - Have these changed at all since 2019?
17. Does your institution treat one-time gifts differently from continuous funding streams?
18. Does your institution conduct classified research?
19. Does your institution work with controlled unclassified information (CUI)?

Campus Participation in the 2019 NDAA Waiver Process

[ASK QUESTIONS 20–23 ONLY IF RESPONDENT PARTICIPATED IN 2019 NDAA WAIVER PROCESS]

20. What is/are the reason(s) that your institution decided to apply for a waiver to host both a DOD Chinese Language Flagship and a Confucius Institute, if applicable? Could you describe this decision?
21. What bottom-line argument did your institution make to the Department of Defense to justify hosting a Confucius Institute and receiving DOD Chinese Language Flagship funding, if applicable?
22. What types of Confucius Institute-related documents and information did your institution provide to the Department of Defense in support of receiving a waiver in 2019?
    - Probe: Did the documents you provided in support of receiving a waiver include agreements, financial records, course syllabi, CI-sponsored speaker series, flyers for cultural events, vision statements, strategic plans, or statements concerning academic freedom?
23. What information and characteristics, if any, does your institution feel demonstrate a justifiable reason for your academic institution to receive a waiver?

Campus Confucius Institute Closure

[ASK QUESTIONS 24–25 ONLY IF CI IS CLOSED]

24. Was your institution's Confucius Institute transferred to another school or organization? Could you describe this decision, if applicable?

25. Following the closure of your Confucius Institute, did your institution create any new organizations, programs, or offerings to continue programs previously conducted or organized by your Confucius Institute or to fill any language and culture programming gaps?
    - If yes, how were these organizations or offerings funded and/or structured?

Concluding Question

26. Is there anything else you would like to share to inform the Confucius Institutes consensus study?